Sustainable Civil Infrastructures

About this Series

Sustainable Infrastructure impacts our well-being and day-to-day lives. The infrastructures we are building today will shape our lives tomorrow. The complex and diverse nature of the impacts due to weather extremes on transportation and civil infrastructures can be seen in our roadways, bridges, and buildings. Extreme summer temperatures, droughts, flash floods, and rising numbers of freeze-thaw cycles pose challenges for civil infrastructure and can endanger public safety. We constantly hear how civil infrastructures need constant attention, preservation, and upgrading. Such improvements and developments would obviously benefit from our desired book series that provide sustainable engineering materials and designs. The economic impact is huge and much research has been conducted worldwide. The future holds many opportunities, not only for researchers in a given country, but also for the worldwide field engineers who apply and implement these technologies. We believe that no approach can succeed if it does not unite the efforts of various engineering disciplines from all over the world under one umbrella to offer a beacon of modern solutions to the global infrastructure. Experts from the various engineering disciplines around the globe will participate in this series, including: Geotechnical, Geological, Geoscience, Petroleum, Structural, Transportation, Bridge, Infrastructure, Energy, Architectural, Chemical and Materials, and other related Engineering disciplines.

More information about this series at http://www.springer.com/series/15140

Don Chen · Sonny Kim · Anand Tapase
Editors

Advancements on Sustainable Civil Infrastructures

Proceedings of the 5th GeoChina International Conference 2018 – Civil Infrastructures Confronting Severe Weathers and Climate Changes: From Failure to Sustainability, held on July 23 to 25, 2018 in HangZhou, China

Springer

Editors
Don Chen
University of North Carolina at Charlotte
Charlotte, NC, USA

Anand Tapase
Karmaveer Bhaurao Patil College of Engineering
Satara, India

Sonny Kim
University of Georgia
Athens, GA, USA

ISSN 2366-3405 ISSN 2366-3413 (electronic)
Sustainable Civil Infrastructures
ISBN 978-3-319-96240-5 ISBN 978-3-319-96241-2 (eBook)
https://doi.org/10.1007/978-3-319-96241-2

Library of Congress Control Number: 2018948644

This Springer imprint is published by the registered company Springer Nature Switzerland AG
The registered company address is: Gewerbestrasse 11, 6330 Cham, Switzerland

Contents

Introduction

For many years, civil infrastructures have mostly been designed and constructed based on direct costs, without proper consideration of the maintenance and depreciation over its service life. However, in recent years much emphasis has been placed on meeting the environmental and socioeconomic targets of sustainable development. This volume constitutes an important contribution that addresses urgent needs to develop sustainable infrastructures, mainly associated with the engineering properties of pavement materials, erosion control, soil mechanics, and pavement performance evaluation. This volume is part of the proceedings of the 5th GeoChina International Conference on Civil Infrastructures Confronting Severe Weathers and Climate Changes: From Failure to Sustainability, HangZhou, China, 2018.

The Use of Automated Flocculation Titrimetry and SARA Analysis to Predict the Performance of Bituminous Binders in Asphaltic Hot Mix and Bitumen Emulsions

Glynn Holleran(✉), Irina Holleran, and Douglas J. Wilson

Department of Civil and Environmental Engineering, The University of Auckland, Auckland, New Zealand
{ghol, imot002}@aucklanduni.ac.nz, dj.wilson@auckland.ac.nz

Abstract. Bitumen is complex chemically, but its composition determines rheology. The Strategic Highway Research Program 1 (SHRP) developed a bitumen microstructural model that relies on the thermodynamic compatibility of the fractions (based on polarity) to create a thermodynamically stable continuum. This stability determines the performance of any bitumen in hot-mix and emulsion applications. Automated Flocculation Titrimetry (AFT) developed by the Western Research Institute (WRI) USA measures Heithaus parameters that define this stability. Chromatographic analysis, known as SARA (saturates, aromatics, resins and asphaltenes) provides a broad compositional analysis that may be used with hiethaus parameters to predict the performance of bitumen in mixes and emulsions. The methodology is based on work from the Western Research Institute at the University of Wyoming Laramie (USA) to measure the known Heithaus (stability) parameters accurately and reproducibly and is detailed in ASTM 6703. In this work, Reclaimed Asphalt Pavement (RAP) binders and virgin binders were examined by SARA analysis and AFT. The properties of the mixes made with these binders are determined and compared to the predictions. A set of emulsion binders with different SARA analysis were also examined for emulsability and the results compared to the AFT results.

1 Introduction

Bitumen is a continuum; the Strategic Highway Research Program 1 (SHRP 1) project carried out in in the USA in the late 1980s by Jones (Jones 1992), and Robertson (1991) developed the microstructural model that describes this behaviour. In this model, the bitumen is a mixture of polar species dispersed in non-polar species. The structural aspect of the binder is provided by interactions between the polar species in the form of pi and hydrogen bonds, as shown in Figs. 1 and 2. Thus the detailed chemistry of the different polarity fractions is important. Hetero species—sulphur, nitrogen, and oxygen provide most of the polar interactions. The basic fractions may be separated by thin layer chromatography (TLC) to separate the bitumen from saturates,

D. Chen et al. (eds.), *Advancements on Sustainable Civil Infrastructures*,
Sustainable Civil Infrastructures, https://doi.org/10.1007/978-3-319-96241-2_1

aromatics, resins and asphaltenes (SARA) analysis as described by Holleran and Holleran (2010). However, this analysis does not reveal the detailed chemistry of individual fractions and hence does not describe the level of internal stability and thus properties fully. The structural properties of the bitumen depend on the balance and the level of associations created by polar interactions. Asphaltenes were once considered to be very large molecules, (Boduszynski 1981) found they are associations of molecules of no more than 4000 Daltons. Agglomerations are thus reversible. In a thermodynamically stable system, the associations are balanced by their dispersion in the oils fraction (saturates and aromatic oils). In the older colloidal model, the concept of peptization or compatibilisation was introduced by Heithaus (1962). Heithaus parameters also adequately assess the microstructural scenario as it is simply a measure of interaction leading to stability. Emulsability is a function of the Hydrophilic-Lipophilic Balance (HLB) of the emulsifier in the bitumen. Asphaltenes are not emulsified but rather dispersed as discussed by Holleran and Holleran (2008). Therefore the dispersion is critical to emulsion stability.

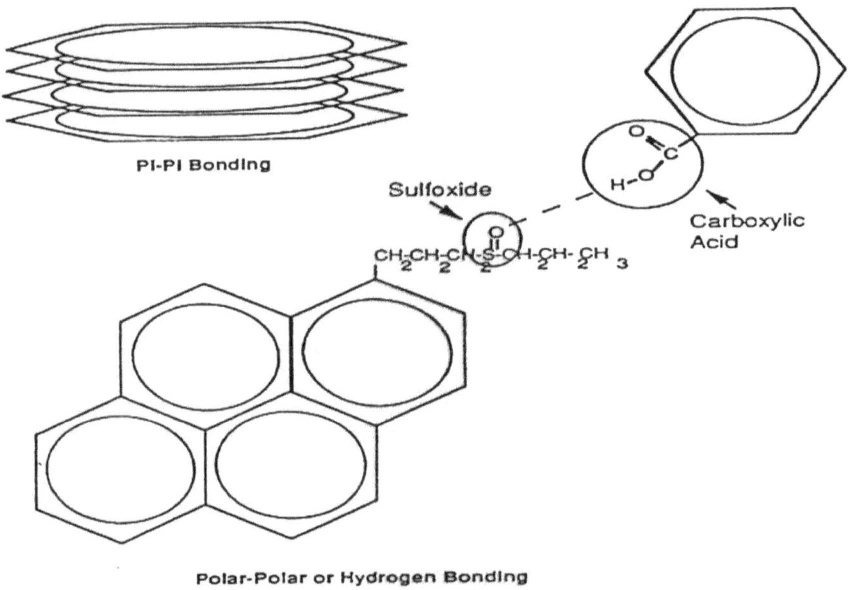

Fig. 1. Primary interactions in bitumen (Holleran and Holleran 2010)

In oxidative aging, SARA fractions are converted by insertion of oxygen into the organic molecules at reactive sites. Aromatics partially converted to resins, resins to asphaltenes, and asphaltenes further increased in polarity (Holleran et al. 2017; Planche 2014; Harnsberger et al. 2008). Hence RAP binders may be thermodynamically unstable depending on the degree of oxidation and its effect on associated species and their level of dispersion. Thus, Heithaus parameters with SARA may be used to assess

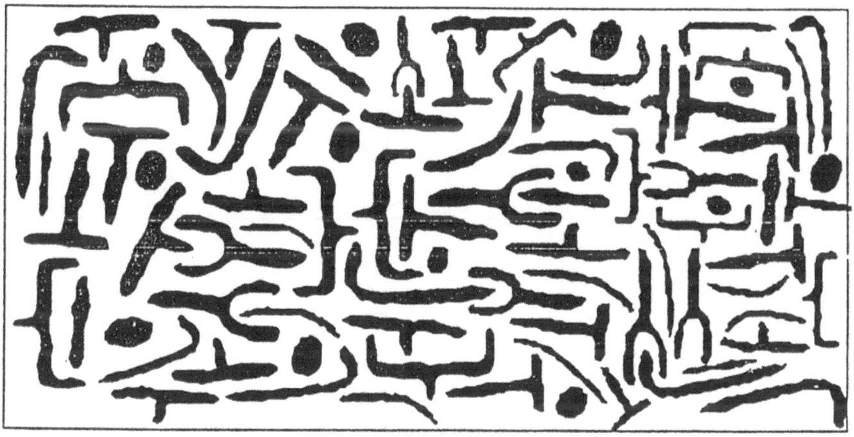

Fig. 2. Microstructural model (Planche 2014)

the condition of the binder, its emulsability, and recyclability of RAP. Rejuvenation effect may also be assessed also by use of Heithaus parameters.

2 Research Objective

The research objective was to verify the use of Heithaus parameters as measured using AFT and the SARA analysis with the physical performance of the binder in RAP mixes and emulsions.

3 Materials Used

The bitumens were both paving grade and emulsion grade bitumen derived from New Zealand Refinery (NZRC) and imported sources (USA). The RAP came from pavement milled after approximately ten years' service life. Extraction was by Dichloromethane and recovery with a centrifuge and final solvent removal using rolling thin film oven at 90 °C under dry nitrogen. The PG grade virgin binder was a PG64-22H made to New Zealand Transport Agency (NZTA) M/1A Specification. The emulsion grades were 80–100 penetration made to NZTA M/1 specification. The rejuvenator was a lignin modified oil as provided by an outside supplier. Mixes were made and tested using NZTA M/10 (2014) methodology and specifications. A standard nominal maximum aggregate size of 14 mm (AC14 as per specification) and a greywacke aggregate source. The grading and binder contents were optimised using NZTA M/10 2014. The difference between mixes was only in the type of binder.

4 Basic Bitumen Chemistry and Sara

Bitumen chemistry is complex and dependent on both crude and refining. The differences in the composition have been demonstrated by Holleran and Holleran (2010). The composition of bitumen is plotted on a triangular diagram that shows the SARA fractions. It is possible to define a range of compositions that have satisfactory rheology. Appropriate blends with RAP, as shown in Fig. 3, can be determined as well as rejuvenation materials and levels.

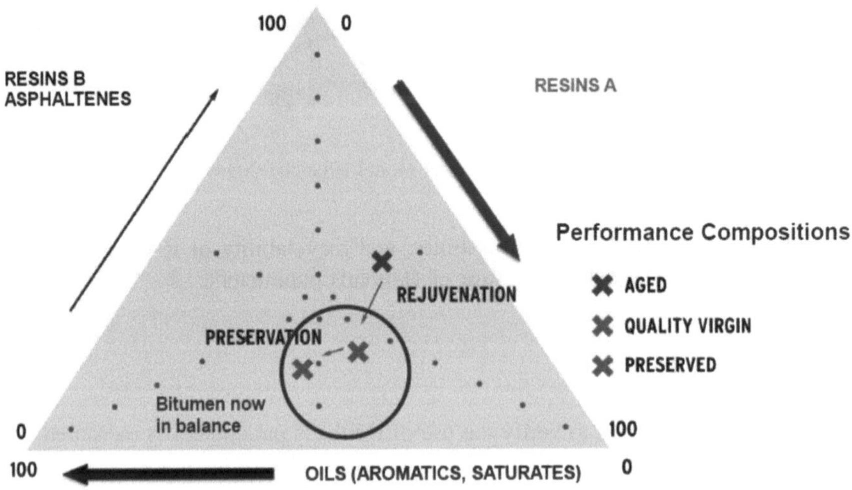

Fig. 3. SARA analysis (Holleran and Holleran 2010)

5 AFT Development and Description

Pauli (1996) has described the development of the Automated Heithaus test and its usefulness in characterising bitumen. The test was originally developed to describe asphaltene chemistry in crude oils, but for nearly four decades it has been used to determine thermodynamic stability in blended bitumens. It is very useful in describing the microstructural changes bitumen goes through with oxidation and refinery processes. The incompatibility referred to, came from the colloidal model which is divided into sol and gel bitumens where the gels are the dispersed medium and sols the dispersing medium. Hence, the Heithaus test is designed to measure this solubility function in polar terms. The gels are micelle structures analogous with emulsion technology. Micelles, cannot be observed in bitumen. There are no phase boundaries in bitumen the microstructural model being based on polar species dispersed in less polar and non-polar species. So, compatibility concerns mutual miscibility, and this determines stability. Oxidation changes this balance as do different heteroatom compounds in both the polar and non-polar phases. Oxidation of bitumen is described by Holleran et al. (2017), and the effect is to destabilise the bitumen microstructure. The methodology has been described by Pauli (1996). It calculates three basic parameters. Namely p_a—peptisability of the associated species, p_o—the peptizing power of the dispersing species, and P—the overall stability of the system. The classical methodology requires manual titration as follows:

- Four 1.0 g samples of a test bitumen are placed into four 125 mL Erlenmeyer flasks.
- To the four flasks are added amounts of 1.0, 2.0, 4.0, and 6.0 mL toluene, respectively.
- After the dissolution of the bitumen is completed, the flasks are immersed in a water bath maintained at 25 °C (77 °F) for 30 min.
- The flasks are titrated with 1.0 mL aliquots of n-heptane. After each addition of n-heptane, the contents of the flask are stirred for several minutes and then inspected to observe if flocculation has taken place. Flocculation is detected by transferring a drop of the p the solution to a filter paper with a glass rod.
- The development of two rings on the filter paper signifies the onset of flocculation.

Heithaus parameters p_a, p_o, and P are calculated as described by Pauli (1996). In ASTM D 6703 (2014) an automated instrument is used shown in Fig. 4.

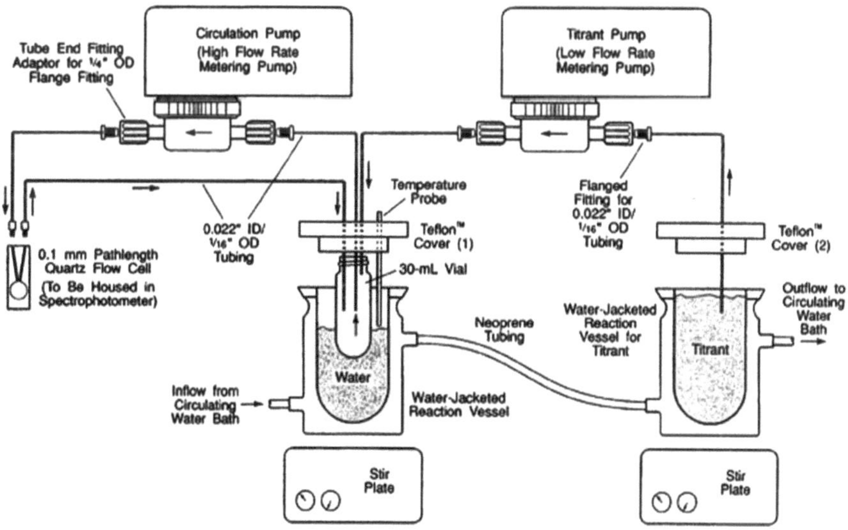

Fig. 4. Heithaus automated schematic (Planche 2014)

- Three solutions with different concentrations are made as follows. Three bitumen samples of weights 0.4, 0.6, and 0.8 g are transferred to separate reaction vials and diluted with 3.000 mL toluene to make the three different bitumen solutions.
- Each solution is titrated with isooctane (2,2,4-trimethylpentane) to allow the onset of flocculation as illustrated in Fig. 4.

- During titration, the solution passes through the spectrophotometer, and the output signal is recorded. The computer records and plots the change of percent transmittance (%T) of detected radiation a (740 nm wavelength) that passes through the quartz cell. This parameter is plotted against the time (*t*) of titration.
- The calculated Flocculation Ratio (FR) versus Concentration (C) graph gives FR_{max} and C_{min} (y and x-intercepts, respectively). These two parameters are used to calculate the Heithaus parameters as shown below:

$$pa = 1 - FR\text{max}$$

– typically ranges from 0.5 to 1.0.

$$po = FR\text{max}(1/C\text{min} + 1)$$

– typically ranges from 0.6 to 1.2.

$$P - po/(1 - pa)$$

– Typically ranges from a highly stable value near 4 to an unstable value near 2.

6 Testing and Results Asphalt Grades

6.1 SARA Analysis

The SARA analysis is presented in Figs. 5 and 6. It shows the compositions of the main binders tested. The RAP binder is very high in resins B and deficient in aromatics, the RAP blends show a proportional adjustment in composition. The rejuvenated binder is not greatly different in its composition after rejuvenation. The Gaestel index (GI), a measure of the colloidal stability (the ratio of the percentage of asphaltenes (Resins) B plus the percentage of Saturates divided by the percentage of resins A, plus the percentage of aromatics), however, shows an increase in colloidal stability after rejuvenation. The data shows that oxidation has increased Resins B, lowered Resins A marginally, and significantly reduced Aromatics. While the GI has not been shown to be a good indicator of thermodynamic, stability the trend is correct.

6.2 AASHTO M332 Grading

The AASHTO M332 (2014) grading of the recovered binders is shown in Table 1. These binders have been graded to the true PG grading using the full AAHSTO M322 (2014) specification. The virgin binder as used in New Zealand has been graded to NZTA M-1A (2016) grade H and is a standard heavy traffic grade with an upper temperature of 64 °C and lower temperature of −22 °C. This grade is designated as PG64-22H. The RAP binder was significantly oxidised. The Rap binder grade is as a high temperature of 88 °C and low temperature of only 0 °C. The RAP grade may be expected to be more prone to cracking. This data is consistent with SARA analysis.

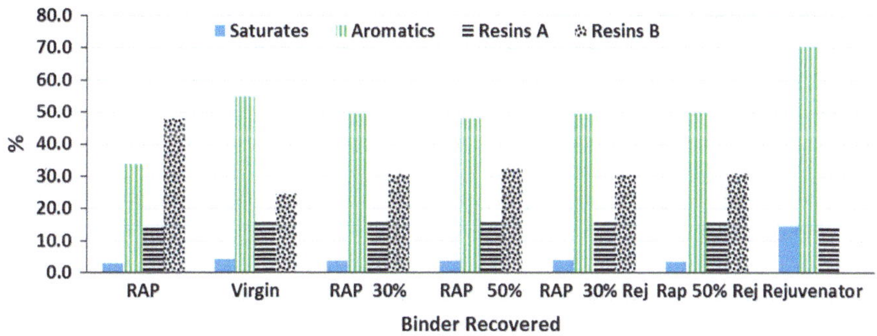

Fig. 5. SARA analysis

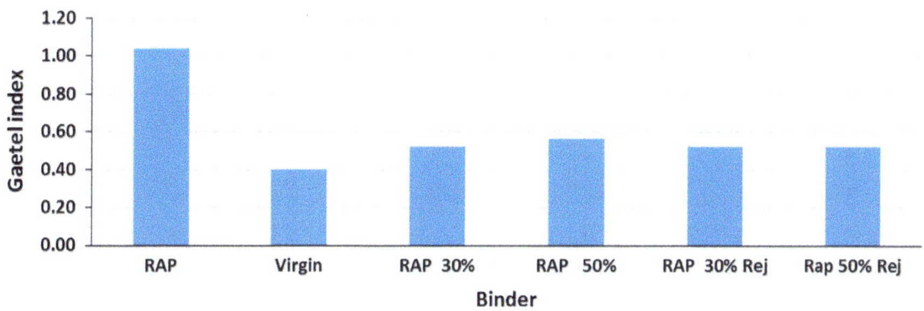

Fig. 6. Gaestel Index

Table 1. Binders grades (true grade)—AASHTO M332 (2014)

Binder	Grade
Virgin	PG64-22 Grade H
RAP	PG88-0 Grade H
30% RAP	PG76-12 Grade H
50% RAP	PG76-6 Grade H
30% RAP rejuvenated	PG70-12 Grade H
50% RAP rejuvenated	PG70-12 Grade H

When combined with the base binder the PG grades are improved at the high-temperature end, but low temperature grading would suggest poor low temperature performance. 50% RAP being worse than 30%. When rejuvenator is added to the aged binder, the low temperature properties significantly improve, and the high-temperature properties move closer to that of the virgin binder.

6.3 Mix Properties

The mix properties were measured on an AC14 NZTA M/10 (2014) heavy traffic mix design.

6.3.1 Wheel Tracking

The wheel tracking was carried out as specified in NZTA (2014) results are shown in Fig. 7. It showed that the RAP consistently improved the deformation resistance and that this was greatest for 50% RAP. The rejuvenator did not change the wheel tracking result much. All the results were satisfactory and in compliance with the specification. The ratio between the deformations at 10,000 and 20,000 cycles show that the system was close to interlock of aggregate, i.e. a flat response.

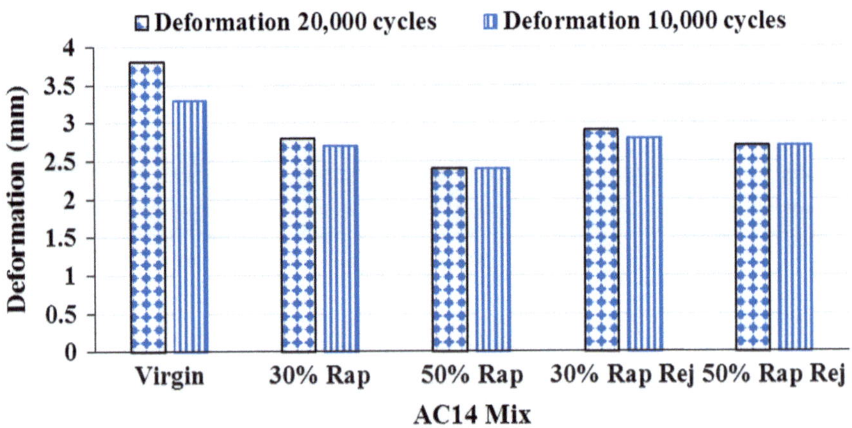

Fig. 7. Wheel tracking

6.3.2 Fatigue

The fatigue properties were tested using 4 pt bending as specified in NZTA (2014) results are shown in Fig. 8. This data shows that the fatigue life of the RAP mixes was lower than that of virgin mixes but that this was improved by use of rejuvenation and is consistent with SARA and rheological testing of binders.

6.3.3 Overlay Testing

The overlay testing was carried out as specified in TXDOT (2017), results are shown in Fig. 9. This data shows that the cracking resistance of the RAP mixes is lower than virgin mixes, but that rejuvenation recovered this to a satisfactory level.

6.3.4 AFT Testing

The AFT was carried out using the approach of Pauli (1986) results are shown in Fig. 10. The peptisability of the asphaltenes and the peptizing power of the dispersing phase show the effect of changes in composition. The Heithaus parameter P shows that the RAP is unstable as a single material and that blending with virgin binder though

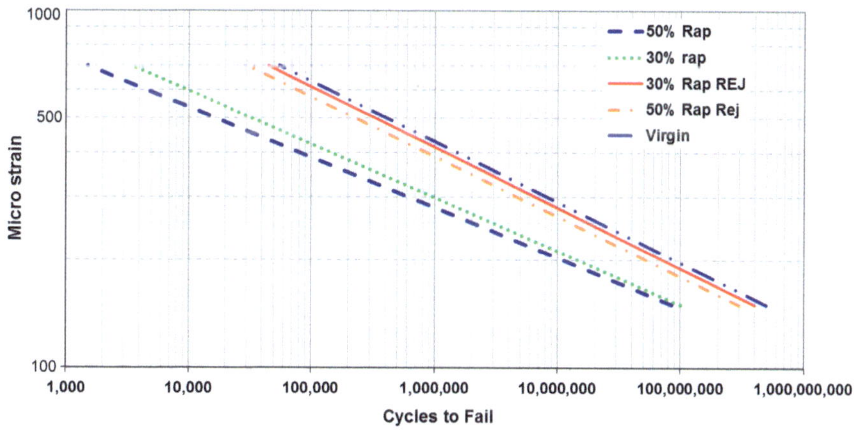

Fig. 8. Fatigue results

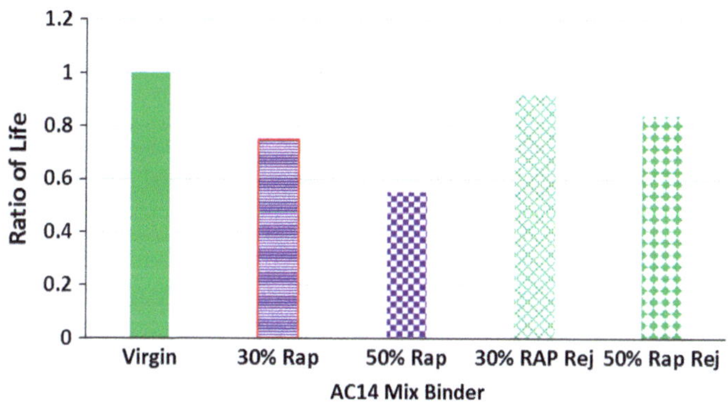

Fig. 9. Overlay test results

improving the stability does not bring it to virgin levels. Rejuvenation improved the stability significantly. The Heithaus parameter would appear to be consistent with the mix results and the SARA analysis.

6.4 Testing and Results Emulsion Grades

Testing for emulsion grades is shown in Tables 2 and 3 and Figs. 11, 12, 13 and 14. The focus was on emulsion stability as tested by sedimentation and coalescence. The presence of storage issues and product performance is highlighted. The ten bitumens represented a range of crude bases and processing methods. The SARA results show a

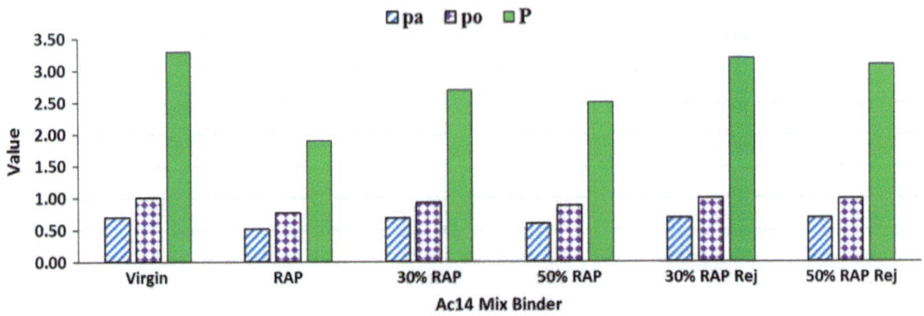

Fig. 10. AFT test results

range of different compositions, the Heithaus parameters, and to a lesser extent, the GI shows that the emulsion performance with a single emulsification system could be

Table 2. Emulsion test results—SARA

Bitumen	Saturates (%)	Aromatics (%)	Resins A (%)	Resins B (%)	Dispersing (%)	Associated (%)	GI
1	4.00	57.30	13.40	25.30	61.30	38.70	0.41
2	5.00	60.90	12.90	21.20	65.90	34.10	0.35
3	4.60	59.20	12.00	24.20	63.80	36.30	0.40
4	6.30	50.30	18.00	25.40	56.60	43.40	0.46
5	4.00	48.90	13.00	34.10	52.90	47.10	0.62
6	4.60	59.20	12.00	24.20	63.80	36.30	0.40
7	4.00	48.90	13.00	34.10	52.90	41.10	0.62
8	6.20	64.20	13.30	16.30	70.50	29.50	0.29
9	4.70	58.60	15.20	21.60	63.30	36.70	0.36
10	6.00	49.00	14.50	30.40	55.00	44.90	0.57

Table 3. Emulsion test results—AFT and performance

Bitumen	p_a	p_o	P	Sedimentation %	Field performance
1	4.00	57.30	13.40	23.5	Poor storage
2	5.00	60.90	12.90	2.5	OK
3	4.60	59.20	12.00	2.2	OK
4	6.30	50.30	18.00	15.2	Slow cure
5	4.00	48.90	13.00	16.5	Slow cure
6	4.60	59.20	12.00	1.8	OK
7	4.00	48.90	13.00	13.5	Clogging spray
8	6.20	64.20	13.30	0.9	OK
9	4.70	58.60	15.20	1.1	OK
10	6.00	49.00	14.50	20.5	Poor storage

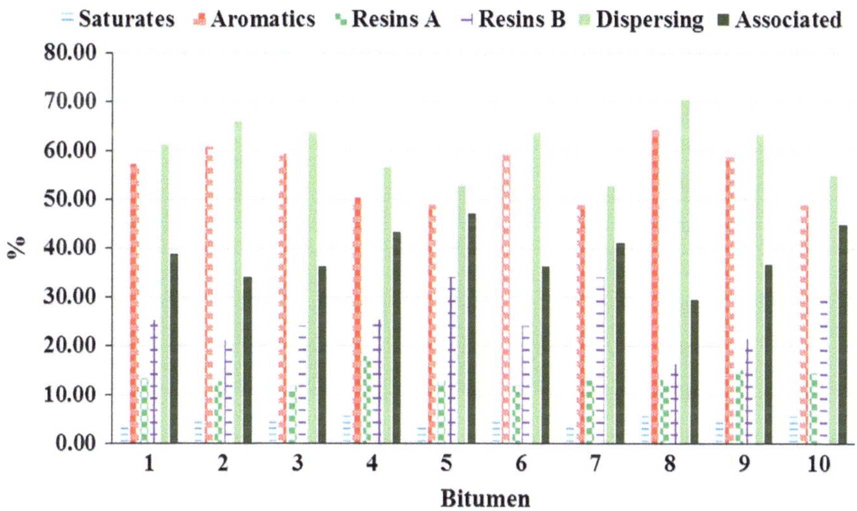

Fig. 11. Emulsion binder—Iatroscan results

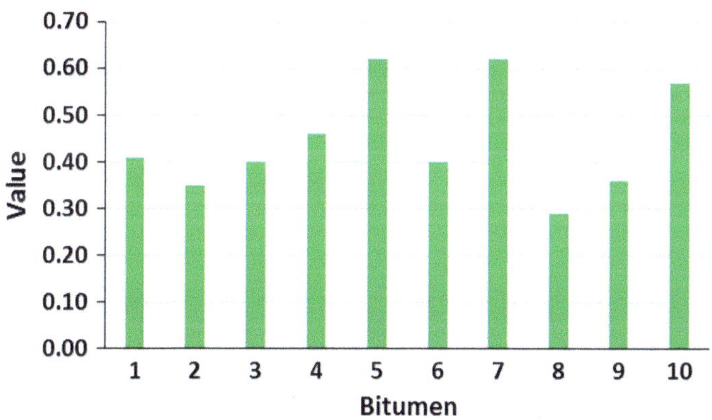

Fig. 12. Emulsion binder—Gaestal Index results

predicted. The field and laboratory testing of cure are in agreement with this prediction. In general, the Heithaus factor P needs to be around 3 or higher. This figure represents a more stringent requirement that appears to be the case in the asphaltic hot mix. More data is needed to set the exact limits.

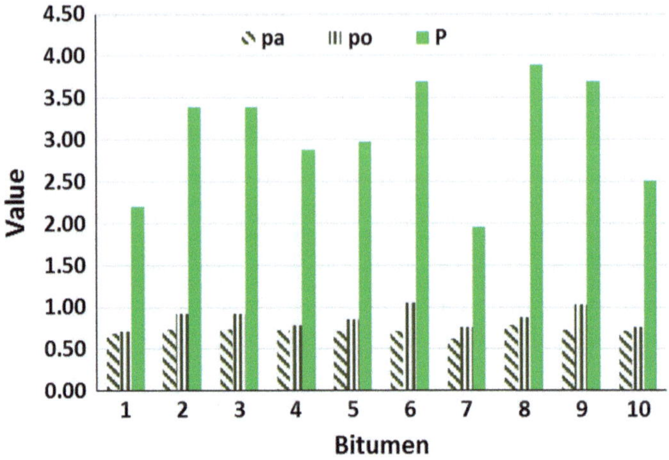

Fig. 13. AFT results

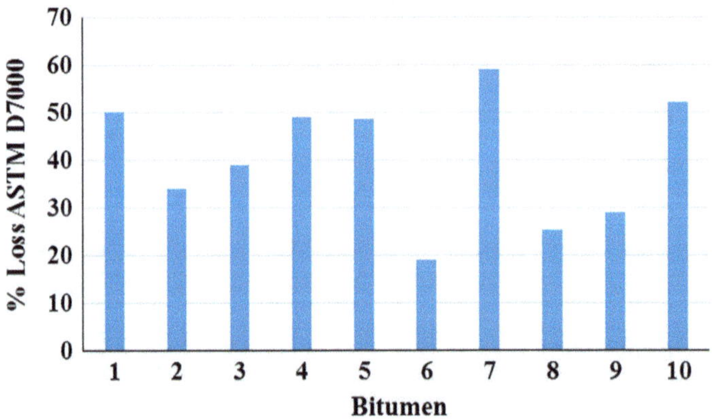

Fig. 14. Sweep test results ASTM D7000 (2017)

7 Conclusions

- The AFT methodology appears to predict the performance of bitumen by determining the agglomeration tendency of the polar fraction of the binder.
- There is broad agreement on AFT parameters with the SARA compositions. The Gaestel Index is in broad agreement with both other sets of data.
- The AFT is correlated with the rheological properties as measured in the PG third generation specification approach.
- RAP binders have a general tendency to higher polarity and hence agglomeration this may be mitigated by blending with more balanced bitumen or by rejuvenation. SARA analysis may be used as a blending tool, but AFT and performance testing must be used to create an accurate prediction of properties.

- Low AFT P factors result in the poorer performance of mixes in crack resistance and low temperature properties and are reflected in low temperature rheological properties in the binder.
- In emulsions, the detailed chemistry is important, but the AFT appears consistent with emulsion properties and the SARA data.
- The AFT is a valuable tool to determine RAP and rejuvenation use in hot mix and performance in emulsions.

References

AASHTO; AASHTO M332 (2014): Standard Specification for Performance-Graded Asphalt Binder Using Multiple Stress Creep Recovery (MSCR) Test. American Association of State Highway and Transportation Officials, Washington, D.C. (2013)

ASTM International: ASTMD 6703: Standard Test Method for Automated Heithaus Titrimetry. ASTM International, West Conshohocken, PA USA (2014)

ASTM International and ASTM D7000-11: Standard Test Method for Sweep Test of Bituminous Emulsion Surface Treatment Samples. ASTM International, West Conshohocken, PA USA (2017)

Boduszynski, M.M.: Asphaltene in petroleum asphalts: composition and formation. In: Advances in Chemistry Series: Chemistry of Asphalts. American Chemical Society, Washington, D.C. (1981)

Harnsberger, M., Peterson, C., Pauli, T.: Asphalt Component Compatibility—RAP. Expert Task Group Meeting, FHWA, Western Research Institute Wyoming (2008)

Heithaus, J.J.: Measurement and significance of asphaltene peptization. J. Inst. Pet. **48** (1962)

Holleran, G., Holleran, I., Bearsley, S., Dubois, C.J., Wilson, D.: Epoxy asphalt for durability of open graded mixes. In: Proceedings of Australian Asphalt Pavement Association, 17th International Conference, Melbourne (2017)

Holleran, G., Holleran, I.: Bitumen chemistry using cheaper sources—an improved method of measurement by TLC-FID and the characterization of bitumen by rheological and compositional means. In: Proceedings Australian Road Research Board, 24th International Conference, Melbourne (2010)

Holleran, G., Holleran, I.: Coalescence control in sealing emulsions. In: Proceedings Australian Road Research Board, 1st Spray Sealing Conference, Adelaide (2008)

Jones, D.R.: An asphalt primer: understanding how the origin and composition of paving-grade asphalt cements affect their performance. SHRP Technical Memorandum #4, National Research Council, Washington, D.C. (1992)

NZ Transport Agency and TNZ M/1A: Performance specification for asphalt binders. NZ Transport Agency, Wellington, New Zealand (2016)

NZ Transport Agency: TNZ M/1 Specification for roading bitumens. NZ Transport Agency, Wellington, New Zealand (2005)

NZ Transport Agency: TNZ M/10 Specification for dense-graded and stone mastic asphalts. NZ Transport Agency, Wellington, New Zealand (2014)

Pauli, A.: Asphalt compatibility by using the automated heithaus titration test. Preprints of papers, Am. Chem. Soc. Div. Fuel Chem. **41** (1996)

Planche, J.P.: Insights into binder chemistry microstructure properties relationship. In: Proceedings of International Society of Asphalt Pavements (ISAP), Raleigh North Carolina (2014)

Robertson, R.E.: Chemical properties of asphalts and their relationship to pavement performance. Report # SHRP-A-UWP-91-510, National Research Council, Washington, D.C. (1991)
Texas Department of Transport: Overlay Test Tex-248-F, Austin, Texas, USA (2017)

Laboratory Characterization of Low-Temperature Performance of High Modulus Asphalt Concrete by Beam Bending Test

Xiaobo Wen[1,2](✉) and Haoran Zhu[1]

[1] JSTI Group Co. Ltd., 2200 Chengxin Road, Nanjing, China
wxbl68@jsti.com
[2] Jiangning High-Tech Industrial Park, 2200 Chengxin Road, Nanjing, China

Abstract. Low-temperature cracking is a serious distress of asphalt pavement which cannot be ignored in cold climates. In this study, the low-temperature performance of high modulus asphalt concrete is characterized through low-temperature beam bending test. According to the analysis of low-temperature cracking mechanism and test results in this paper, the critical value of strain energy density is a reasonable index to evaluate the low-temperature performance which combines the critical bending strength and strain. Results also show that the dosage of high modulus additives has a significant influence on the low-temperature performance of asphalt concrete. Road engineers should keep a balance between the high-temperature performance and low-temperature performance of high modulus asphalt concrete, especially in some cold regions.

Keywords: Highway engineering · High modulus asphalt concrete
Beam bending test · Low-temperature performance · Strain energy dense

1 Introduction

Rutting is one of the most important distresses for asphalt pavements [1]. It is caused by material consolidation and lateral movement due to repeated heavy wheel loadings on the various pavement layers. The rutting distress is viewed as not only a structure failure but also a serious safety hazard to vehicles because hydroplaning can occur in the presence of rutting in rainy weather, resulting in serious traffic accidents. Moreover, vehicles tend to be pulled towards the rutted path, making it uncomfortable to drive [2, 3]. Researchers have put forward many solutions to the rutting problem [4]. The concept of high modulus asphalt concrete (HMAC) [5] always suffers the attention. By improving the modulus of asphalt concrete, it can decrease the plastic deformation of asphalt concrete under traffic loadings and increase the rutting resistance of asphalt concrete. Therefore, HMAC is an effective way to extend the service life of asphalt pavement. So far, there are two main methods [6] to increase the modulus of asphalt concrete. One is to select low penetration asphalt, which has a higher viscosity. The other is to blend modifiers together with the asphalt concrete. In this study, the second method was adopted.

D. Chen et al. (eds.), *Advancements on Sustainable Civil Infrastructures*,
Sustainable Civil Infrastructures, https://doi.org/10.1007/978-3-319-96241-2_2

A lot of studies [7] has concentrated on the high-temperature performance of HMAC, but low-temperature cracking is a serious distress of asphalt pavement which cannot be ignored in cold regions. Exposed to the fluctuating temperature, both surface and underlying part of the pavement will experience the action of expansion or contraction due to the heat transfer in the pavement. Low-temperature cracking [8] induced by seasonal and daily thermal cyclic loading is a significant cause of premature pavement deterioration and quite harmful to the service life and quality of the road. Currently, fewer works [9] have involved the low-temperature performance of HMAC. To understand the behavior of HMAC at low service temperatures and predict their field performance, a constitutive stress-strain relationship must be described. For this purpose, many models were established [10, 11].

2 Objective

Based on the above considerations, the objective of this paper is to characterize the low-temperature performance of HMAC through low-temperature beam bending test. In addition, an index from the low-temperature beam bending test which can reflect low-temperature performance of HMAC accurately will be proposed.

3 Analysis of Low-Temperature Cracking Mechanism

Asphalt mixtures can be seen as elastic materials at low temperature, the destroy process of which is accompanied with energy dissipation. The work external force does to materials can be transformed into two parts: elastic strain energy to be stored and surface energy to produce new interface after initiation and propagation of cracks. The low-temperature cracking of asphalt pavement is [12] due to that internal fatigue stress and strain induced by the thermal stress of asphalt mixture exceeds the ultimate stress and the ultimate strain. Therefore, single stress or strain indices are unreasonable to evaluate the low-temperature anti-cracking performance of asphalt mixture. Generally, asphalt mixture has better low-temperature anti-cracking performance with more elastic strain energy stored. The damage of material can be expressed with strain energy density function *dW/dV* supposed that the destroy form of material is corresponding to energy condition per unit volume. *dW/dV* is shown as Eq. 1.

$$dW/dV = \int_{0}^{\varepsilon_0} \sigma_{ij} d\varepsilon_{ij} \quad (1)$$

In which, *dW/dV* is strain energy density function; σ_{ij} and ε_{ij} are stress and strain component respectively; ε_0 is strain corresponding to peak stress (also called critical strain). The critical value of strain energy density is the area under stress-strain curves when specimen fractures, which can be measured by test like beam bending test, uniaxial compression test and so on.

4 Materials and Sample Fabrication

4.1 High Modulus Additives

Currently, there are many polyolefin additives used to increase the modulus of asphalt concrete, like PR series in France, Bonifiber in America, Dolanit in German and ZQ series in China. In this study, a high modulus additive namely ZQ manufactured by Jiangsu Baoli Asphalt Co. Ltd. in China was blended with the asphalt and aggregate mix directly. ZQ is processed with multiple polymers and resin through special technique, which is a black granular material with a diameter less than 3 mm, as shown in Fig. 1. Its density is 1000–1200 kg/m^3 and softening point is greater than 160 °C. The specific components are listed in Table 1. ZQ is thrown into the asphalt and aggregate mix directly without reaction, shearing and growth process comparing with traditional polymer modified asphalt mixture.

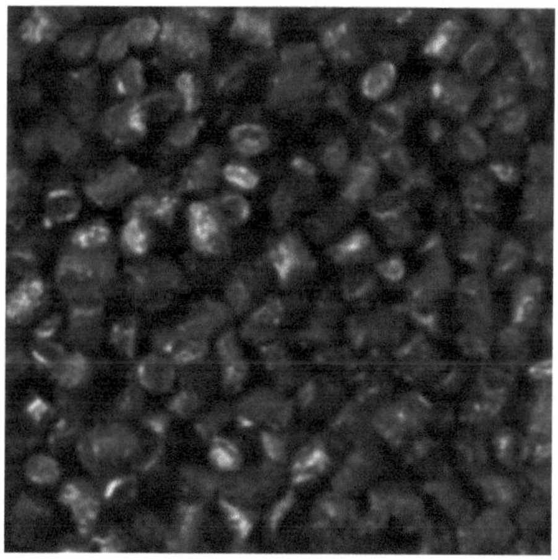

Fig. 1. Photo of ZQ

4.2 Asphalt Binder and Aggregate

The base asphalt binder used in this study is SHELL-70. The basic properties of the binder are listed in Table 2. In this study, basalt aggregate and limestone filler manufactured by Jiangsu Maodi Co. Ltd. in China were used, the properties of which meet the requirements of "Test method of Aggregate for Highway Engineering" (JTG E42-2005).

4.3 Gradation, Mix Design and Specimen Fabrication

Figure 2 shows the recommended gradation limits by Chinese specifications "Technical Specification for Construction of Highway Asphalt Pavements" (JTG F40-2004) for dense graded asphalt mixture (AC-16) with nominal maximum aggregate size (NMAS) of 16 mm.

Table 1. The components of ZQ

Component	Weigh percentage
Polymer	60–80
Cellulose	5–10
Ethylene Vinyl Acetate Copolymer	2–3
Styrene-butadiene rubber (SBR)	20–30
Functional agent	2
Antioxidant RD	0.2

Table 2. Properties of SHELL-70# asphalt

Properties	Value
Penetration (25 °C, 100 g, 5 s, 0.1 mm)	71
Ductility (5 cm/min, 15 °C, cm)	152
Softening point (R&B, °C)	47.5
Flash point (°C)	272
Rotational viscosity (60 °C, Pa.s)	203
Wax content (%)	1.6
Density (15 °C, g/cm^3)	1.032

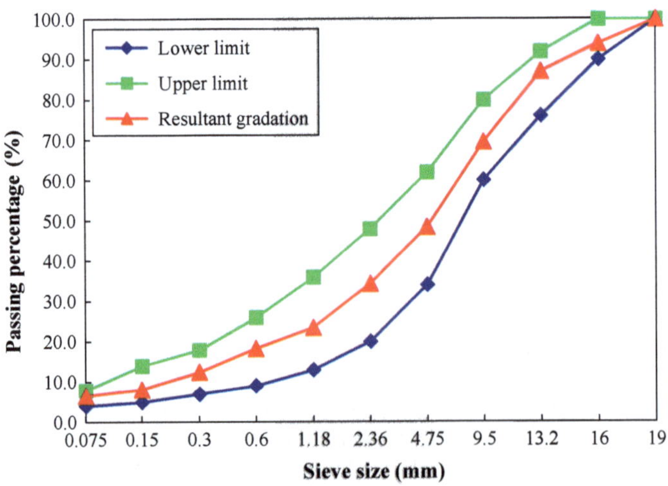

Fig. 2. Curve of aggregate gradation

The mix design procedures for AC-16 in this paper were determined as per the Chinese specification "Standard Test Methods of Bitumen and Bituminous Mixtures for Highway Engineering" (JTG E20-2011). The optimal asphalt content of HMAC is 4.9% (here the asphalt content includes both asphalt and additives), while the basic asphalt mixture is 4.8% based on the Marshall Design method. All the asphalt binders used were SHELL-70. On the first stage, high modulus additives were dispersed and melting into the aggregates. Then, hot asphalt binders were poured into the mix, resulting in the partly dissolution of additives. The obtained mixes are coated by compound binders, in which additives offer strong mechanistic properties and asphalt acts as plasticizer, filler and binder. Three different percentages of ZQ were chosen to produce high modulus asphalt mixtures (5, 7.5 and 10.5% by weight of the base binder). There are four types of asphalt mixtures tested in this study totally.

The mixing and compaction temperatures are 170 and 155 °C, respectively, which were determined using the viscosity-temperature relationship curves. The HMA mixtures were then aged for 2 h at 160 °C in a short-term oven before compaction. The test samples were cut from the square slab specimen of 300 mm × 300 mm × 50 mm compacted using a wheel roller. The sample geometry is 250 mm × 30 mm × 35 mm. The specimen fabrication is according to Chinese specification "Standard Test Methods of Bitumen and Bituminous Mixtures for Highway Engineering" (JTG E20-2011).

5 Experimental Investigation

Currently, there are many testing methods to investigate the low-temperature performance of asphalt concrete, such as the Indirect Tensile Test (IDT), the Thermal Stress Restraint Specimen Test (TSRST), J-integral test, direct tensile relaxation test and so on. Comparing with the above test methods, three-point beam bending test at low temperature is simple and feasible, and the discreteness of test data is lower. Therefore, the low-temperature beam bending test was chosen to evaluate the HMAC low-temperature anti-cracking performance. The schematic of beam bending test is shown in Fig. 3.

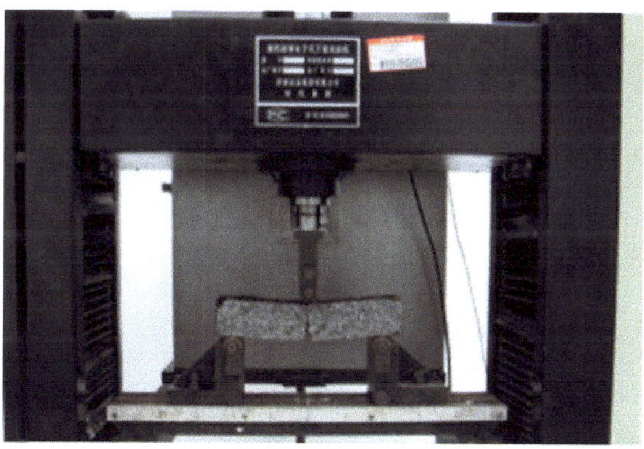

Fig. 3. The schematic of beam bending test

Beam bending test (three-point bending test) was carried out as per Chinese specification with Universal Testing System (UTM-25). Through this test bending strength and bending strain can be obtained, its calculation equations are shown as follows.

$$R_B = 3LP_B/2bh^2 \tag{2}$$

$$\varepsilon_B = 6hd/L^2 \tag{3}$$

$$S_B = R_B/\varepsilon_B \tag{4}$$

In which, R_B is bending strength (MPa); ε_B is bending strain ($\mu\varepsilon$); P_B is peak of load (N); S_B is bending stiffness (MPa); d is the deflection when specimen destroyed (mm); L is the span of the specimen; b is width of the span centre section (mm); h is height of the span centre section (mm).

In this study, all the beam specimens were placed inside the chamber at −10 °C for 1 h using a universal testing machine and temperature control cabinet before testing. Afterward, they were tested at the temperature of −10 °C and loading rate of 50 mm/min. Each type of asphalt mixture has two replicates.

6 Results and Discussion

6.1 Beam Bending Test Results

Four different asphalt mixtures were tested as per Chinese specification at −10 °C with Universal Testing System (UTM-25). The stress-strain curves are shown in Fig. 4.

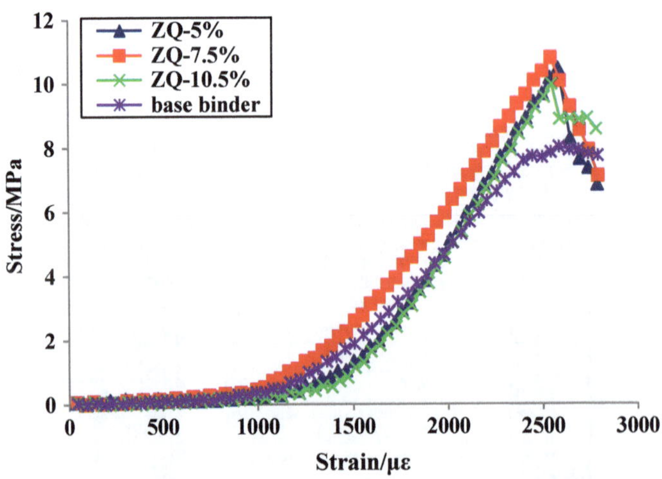

Fig. 4. The stress-strain curves of different asphalt mixtures

From Fig. 4 it can be found that the stress-strain curve of HMAC with 5% ZQ is similar to the one with 10.5%. Comparatively, the stress-strain curve of HMAC with 7.5% ZQ is higher, while the stress-strain curve of basic asphalt concrete is lower. It is inferred that different stress-strain curve means different low-temperature performance.

6.2 Low-Temperature Performance Evaluation Indices

To understand the low-temperature anti-cracking performance of different asphalt mixtures quantitatively, the beam bending test results are shown in Table 3. The strain energy density of asphalt mixture can be obtained from beam bending test results. The area under the strain-stress curve when the beam is a failure is the bending strain energy density of asphalt mixture.

Table 3. Beam bending test results at −10 °C

Mixture type	R_B/MPa	ε_B/ με	S_B/MPa	dW/dV/(kJ m^{-3})
Base binder	8.026	2586	3103.503	6.201
ZQ-5%	10.473	2581	4057.979	6.454
ZQ-7.5%	10.805	2542	4250.245	7.578
ZQ-10.5%	9.983	2547	3918.837	5.762

In order to compare different low-temperature performance evaluation indices more distinctly, the normalized value of critical bending strength, bending strain and strain energy density taking basic asphalt mixture as a reference is curved in Fig. 5.

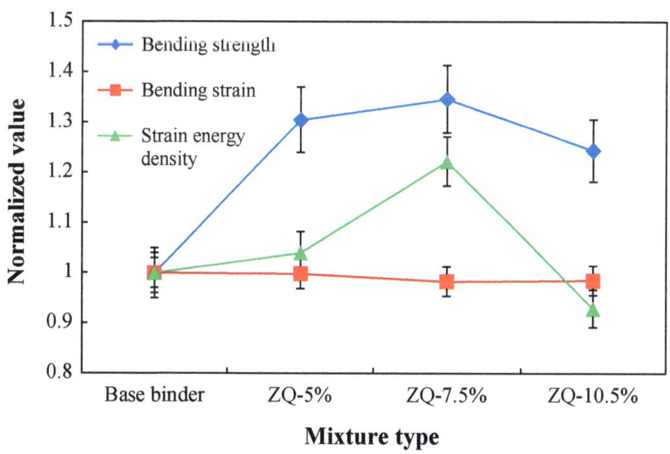

Fig. 5. Different low-temperature performance evaluation indices

It can be concluded from Fig. 5 that three low-temperature performance evaluation indices have different changing trends. The ranking of three evaluation indices on the low-temperature performance of four kinds of asphalt mixtures even appears to be contradictory. It is because the low-temperature performance of asphalt mixture determined by many factors, not only determined by strength characteristics but also deformation capacity, any one of them is inappropriate to evaluate asphalt low-temperature performance. As mentioned in the low-temperature cracking mechanism, according to the material damage principle, the cracking is due to the energy accumulated greater than the critical value, so the strain energy density of asphalt mixture is much better than other and adopted to evaluate the low-temperature performance in this study.

6.3 Influence of the Dosage of ZQ on the Low-Temperature Performance of HMAC

The greater bending strain energy density, the better low-temperature performance asphalt mixture has. The critical values of bending strain energy density of the above four kinds of asphalt mixture are shown in Table 2. From both Table 2 and Fig. 4, it can be found that with the increase of ZQ dosage from 5 to 7.5%, the low-temperature performance of HMAC elevates, and both excel basic asphalt concrete. However, the bending strain energy density drops a lot when the dosage of ZQ is up to 10.5%, even lower than basic asphalt mixture. According to former research [13], increasing the dosage of ZQ can improve the high-temperature performance and rutting resistance of HMAC significantly. The addition of 10.5% ZQ improves the rutting performance most. Based on this study, it is meaningful to be emphasized that road engineers should keep a balance between the high-temperature performance and low-temperature performance of HMAC, especially in some cold regions.

7 Conclusions

Based on the laboratory test results and analysis made in this research study, the following conclusions can be obtained.

1. The low-temperature property evaluated by the indices of critical bending strength and strain are not consistent, it is because the low-temperature performance of asphalt mixture is determined by many factors, not only determined by strength characteristics or deformation capacity.
2. The critical value of strain energy density is a reasonable index to evaluate the low-temperature performance which combines the critical bending strength and strain.
3. With the increase of ZQ dosage from 5 to 7.5%, the low-temperature performance of HMAC elevates, and both excel base asphalt concrete. But the low-temperature performance of HMAC with 10.5% ZQ is inferior to basic asphalt concrete.
4. Road engineers should keep a balance between the high-temperature performance and low-temperature performance of HMAC, especially in some cold regions.

References

1. Wang, H., Yang, J., Liao, H., et al.: Electrical and mechanical properties of asphalt concrete containing conductive fibers and fillers. Constr. Build. Mater. **122**, 184–190 (2016)
2. Wang, H., Jun, Y., Gong, M., et al.: Laboratory performance evaluation of high-modulus asphalt mixtures for perpetual asphalt pavements. In: Transportation Research Board 94th Annual Meeting (15-2614) (2015)
3. Zhao, W., Xiao, F., Amirkhanian, S.N., et al.: Characterization of rutting performance of warm additive modified asphalt mixtures. Constr. Build. Mater. **31**, 265–272 (2012)
4. Taher, B.M., Mohamed, R.K., Mahrez, A.: A review on fatigue and rutting performance of asphalt mixes. Sci. Res. Essays **6**(4), 670–682 (2011)
5. Lee, Hj, Lee, J.H., Park, H.M.: Performance evaluation of high modulus asphalt mixtures for long life asphalt pavements. Constr. Build. Mater. **21**(5), 1079–1087 (2007)
6. Von Quintus, H.L.: Hot-mix asphalt layer thickness designs for longer-life bituminous pavements. Transp. Res. Circular **503**, 66–78 (2001)
7. Yang, P.: Characteristic Behavior of High Modulus Asphalt and High Modulus Asphalt Mixtures. South China University of Technology, Guangzhou (2012)
8. Alavi, M.Z., Hajj, E.Y., Morian, N.E., et al.: Low temperature characterization of asphalt mixtures by measuring visco-elastic properties under thermal loading. In: ISCORD 2013@ sPlanning for Sustainable Cold Regions. ASCE, pp. 404–415 (2013)
9. Behnia, B., Dave, E.V., Ahmed, S., et al.: Effects of recycled asphalt pavement amounts on low-temperature cracking performance of asphalt mixtures using acoustic emissions. Transp. Res. Rec.: J. Transp. Res. Board **2208**(1), 64–71 (2011)
10. Selvadurai, A.P.S., Au, M.C., Phang, W.A.: Modelling of low-temperature behaviour of cracks in asphalt pavement structures. Can. J. Civ. Eng. **17**(5), 844–858 (1990)
11. Kim, H., Buttlar, W.G.: Finite element cohesive fracture modeling of airport pavements at low temperatures. Cold Reg. Sci. Technol. **57**(2), 123–130 (2009)
12. Marasteanu, M.O., Zofka, A., Turos, M., et al.: Investigation of Low Temperature Cracking in Asphalt Pavements: National Pooled Fund Study 776 (2007)
13. Shi, X., Yang, J., Chen, X.: Research on high and low temperature performance of high modulus asphalt mixtures. J. China Foreign Highw. **33**(006), 309–312 (2013)

Author Biography

Xiaobo Wen received his Master degree in transportation engineer from Southeast University, Nanjing, China, in 2016, and then he entered JSTI Group engaged in pavement technology research and development till now.

His current research interests include asphalt pavement preservation scheme design, maitenance materials R&D and sustainable development technology.

Numerical Analysis at the Particulate Scale of Sand Erosion Under Impinging Jet

Kun Zhang[1(✉)], John Petrie[2], Soo-Min Ham[3], Balasingam Muhunthan[1], and Tae-Hyuk Kwon[3]

[1] Department of Civil and Environmental Engineering, Washington State University, Pullman, WA 99163, USA
{kun.zhang2,muhuntha}@wsu.edu

[2] Civil Engineering Department, California State University, Los Angeles, CA 90032, USA
jpetrie@calstatela.edu

[3] Department of Civil and Environmental Engineering, Korea Advanced Institute of Science and Technology, Yuseong-Gu, Daejeon 34141, Republic of Korea
{soomin.ham,t.kwon}@kaist.ac.kr

Abstract. The impinging jet test is widely used in laboratory and in situ to determine critical shear stress and erodibility rate of soils. However, quantification of the forces under this test at the soil particulate level is complex and results in varying interpretation of measured data. The computational fluid dynamics (CFD) has shown effectiveness to study the jet flow features and its effects on erosion. But it cannot explicitly account for the interaction forces between the fluid and soil particles during the dynamic erosion process. This study employs a methodology of the coupled CFD and discrete element method (DEM) to account for such interaction forces directly to study the sand erosion mechanisms under an impinging jet. The interaction forces, including the Di Felice drag force, pressure gradient force, viscous force, and buoyant force, between the fluid and sand particles were incorporated and analyzed during the sand erosion process. A total of nine cases with varying jet Reynolds numbers and ratios of jet height to nozzle diameter (H/D) were simulated. The results show that the coupled CFD and DEM analysis can capture the sand erosion features and scour patterns well, which are dependent on the ratios of H/D and jet Reynolds numbers. By examining the evolution of the maximum fluid-particle interaction forces, different erosion stages, including scour onset, scour developing, and scour stabilization, are identified. Finally, the distributions of the fluid-particle interaction forces along the surface of the sand bed provide a better understanding of the formations of different shapes of scour hole and sand erosion mechanisms at the particulate scale.

1 Introduction

The impinging jet erosion test has been increasingly employed to characterize soil erodibility in laboratory and in situ due to its portability and flexibility (Hanson and Cook 2004; Hanson and Hunt 2006). Studies have been carried out to analyze jet flow

D. Chen et al. (eds.), *Advancements on Sustainable Civil Infrastructures*,
Sustainable Civil Infrastructures, https://doi.org/10.1007/978-3-319-96241-2_3

characteristics, evaluate critical shear stress, and develop empirical or semi-empirical mathematic models to predict the soil erosion rate under the impinging jet (Hanson et al. 1990; Mercier et al. 2012; Mercier 2013). However, the fundamental mechanisms of soil erosion under an impinging jet by considering the interaction between fluid and soil particles are not well-understood.

To better understand jet flow and associated soil erosion, numerical analysis, such as computational fluid dynamics (CFD), has been used (Mercier et al. 2012; Weidner et al. 2012; Mercier 2013; Serrano 2015). The CFD analysis has shown effectiveness to characterize the jet flow and determine the maximum shear stress (Weidner 2012), but it cannot explicitly account for the interaction between the fluid and soil particles at the particulate scale during the dynamic erosion process. To directly account for the interaction between fluid and particles, the coupled CFD and discrete element method (DEM) was explored to study soil erosion by Kuang et al. (2013), Zhang et al. (2015), Zhao and Shan (2013) and Chen et al. (2015). However, the soil erosion mechanisms produced by an impinging jet have not been studied at the particulate scale via coupled CFD and DEM analysis. Thus, the objective of this study is to employ coupled CFD and DEM to analyze the sand erosion process under a submerged turbulent impinging jet in water. As part of this analysis, the fluid-particles interaction forces were analyzed and evaluated at different stages of the erosion process to provide a better understanding of the sand erosion mechanisms at the particulate scale under an impinging jet.

In this context, the CFD analysis was first set up and implemented individually to characterize the fluid flow features of the impinging jet. Three different ratios of jet height to nozzle diameter (H/D) and three different values of jet Reynolds number (R_{eo}) were simulated. Subsequently, the coupled CFD and DEM analysis was established to simulate sand erosion process under the impinging jet test. The scour profiles, the maximum scour depth, and the evolutions of the maximum fluid-particle interaction forces were analyzed. Finally, the distribution of the interaction forces at the surface of sand bed was discussed for the impinging jets with different ratios of H/D to understand the erosion mechanisms at the particulate scale.

2 Numerical Models and Setup

2.1 CFD Analysis of Impinging Jet Flow

The purpose of this part is to develop a CFD simulation to capture the key features of the turbulent flow of the impinging jet. This step is necessary before implementing a reliable coupled CFD and DEM analysis with the confidence of jet flow behaviors. An open source CFD software, OpenFOAM, was used to solve the Reynolds-averaged Navier-Stokes equations, which are the governing equations of the impinging jet flow. The k-omega Shear-Stress Transport (k-ω SST) model was applied to capture the turbulent flow developed in the jet domain and analyze the fluid flow near the walls. The solver, pisoFoam, was selected to solve this transient incompressible flow problem (OpenFOAM-3.0.1 2015).

The computational domain was based on the standard jet device that consists of an impinging jet along the centerline of a cylindrical submergence tank (Hanson and Cook

2004). As shown in Fig. 1a, a quarter of the submergence tank was set up by taking advantage of its geometric symmetry. The nozzle diameter (D) is 0.00625 m. The different ratios of H/D were achieved by varying the jet height (H) from the nozzle to the original sand bed, as shown in Table 1. The nozzle inlet velocities were 1, 3, and 5 m/s for different cases of H/D. And the jet Reynolds number for each case was calculated as:

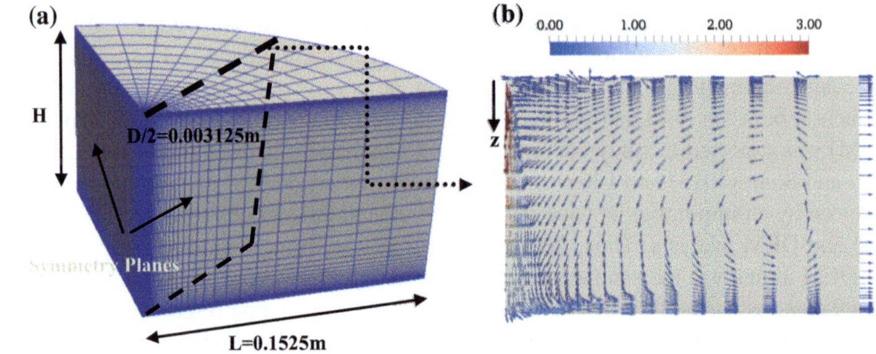

Fig. 1. **a** CFD analysis for Jet flow set-up and **b** fluid velocity vector field at the middle cross-section plane at 15 s

$$R_{eo} = \frac{U_0 D}{v} \tag{1}$$

where R_{eo} is the jet Reynolds number; U_0 is the nozzle inlet velocity; D is the nozzle diameter; and v is the kinematic viscosity of water.

A gradient mesh was created for the CFD domain to capture the near wall fluid velocity, as shown in Fig. 1a. And Fig. 1b presents the jet flow velocity field of the middle cross-section plane at 15 s after the impinging jet begins for the case 5.

2.2 Coupled CFD and DEM Analysis of Sand Erosion by an Impinging Jet

As mentioned previously, the CFD analysis cannot explicitly account for the interaction between fluid and sand particles during the erosion process. Thus, the coupled CFD and DEM analysis was proposed and set up to simulate the sand erosion under an impinging jet in this study. The governing equations of the DEM for the translational and rotational motions of sand particles are expressed as (Zhou et al. 2010):

$$m_i \frac{dv_i}{dt} = \sum_j F_{ij}^n + \sum_j F_{ij}^t + F_i^{pf} + F_i^g \tag{2}$$

Table 1. Parameters for jet flow characterization studies

Case number	D (m)	H (m)	H/D	U_0 (m/s)	R_{eo}	k inlet (m^2/s^2)	ω inlet (1/s)
1	0.00625	0.05	8	1	6250	0.00432	277
2				3	18,750	0.0295	724
3				5	31,250	0.0722	1131
4		0.10	16	1	6250	0.00432	277
5				3	18,750	0.0295	724
6				5	31,250	0.0722	1131
7		0.15	24	1	6250	0.00432	277
8				3	18,750	0.0295	724
9				5	31,250	0.0722	1131

$$I_i \frac{d\omega_i}{dt} = r_i \times \sum_j F_{ij}^t + T_i \quad (3)$$

where m_i and r_i are the particle mass and radius; v_i and ω_i are the translational and rotational velocities of a particle; I_i is the particle moment of inertia; and t is time. The forces acting on the sand particles include: normal contact force (F_{ij}^n), tangential contact force (F_{ij}^t), gravitational force (F_i^g), rolling friction torque (T_i), and particle-fluid interaction forces (F_i^{pf}). The particle-fluid interaction forces are the critical triggering forces that cause the erosion of the sand under the jet flow. In this study, the particle-fluid interaction forces involved the Di Felice drag force, pressure gradient force, viscous force, and buoyant force. The details of the mathematical expressions of these forces can be seen in Zhou et al. (2010).

When the sand particles exist within the fluid domain, the effects of their volumes and the fluid-particle interaction forces lead to the following modification of the Navier-Stokes equations (Zhou et al. 2010):

$$\frac{\partial \alpha_f}{\partial t} + \nabla \cdot \left(\alpha_f \boldsymbol{u}_f\right) = 0 \quad (4)$$

$$\frac{\partial\left(\alpha_f \rho_f \boldsymbol{u}_f\right)}{\partial t} + \nabla \cdot \left(\alpha_f \rho_f \boldsymbol{u}_f \boldsymbol{u}_f\right) = -\alpha_f \nabla p - \boldsymbol{F}_i^{pf}/\Delta V + \nabla \cdot \alpha_f \boldsymbol{\tau} + \alpha_f \rho_f \boldsymbol{g} \quad (5)$$

where α_f is the volume fraction occupied by the fluid; ρ_f is the fluid density; $\boldsymbol{u}_f$ is fluid velocity; p is the fluid pressure; F_i^{pf} is the fluid-particle interaction forces; ΔV is the volume of the CFD cell; and $\boldsymbol{\tau}$ = fluid shear stress.

The turbulence model, k-ω SST, used in the CFD analysis was also used in the coupled CFD and DEM analysis. The solver, cfdemSolverPiso, developed by DSC Computing GmbH (CFDEM® project 2017) was used to perform the coupled CFD and DEM analysis. The set-up of the sand erosion under the impinging jet is shown in Fig. 2. The sand bed thickness (h) is 0.05 m. The sand particle radius is 0.001 m and its density is 2650 kg/m^3. The coefficients of restitution, friction, and rolling friction are input as 0.75, 0.2, and 0.05, respectively, based on the angle of repose of 30° and the Generic Material Model database from the EDEM™ (GEMM 2017). The DEM time step is 0.00001 s and the CFD time step is 0.001 s with the coupling intervals of 100 steps. Similar to the CFD analysis, a total of nine cases were run for the different ratios of H/D and jet Reynolds numbers shown in Table 1. The sand erosion simulation lasts for 15 s in total.

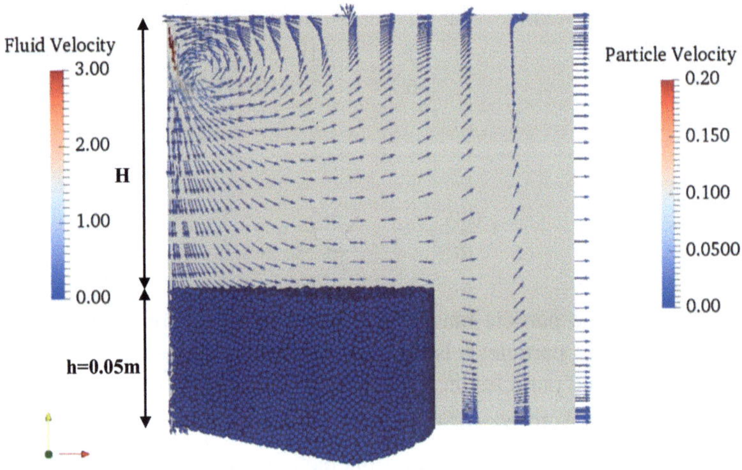

Fig. 2. The set-up of the coupled CFD-DEM analysis for sand erosion simulation under the impinging jet

3 Results and Discussions

3.1 CFD Analysis of Impinging Jet Flow

The profiles of centerline velocity and radial velocity in dimensionless form of the simulated impinging jet are shown in Fig. 3a, b, respectively. As the ratio of H/D increases, the profiles of centerline velocity changes significantly. At the H/D = 8, the centerline velocity decreases drastically right above the stagnation point. When the H/D is higher than 8, the jet flow is fully developed including both free jet region and impingement region along the centerline. However, the jet Reynolds numbers seem to have no significant influence on the dimensionless centerline velocity for the jet with the same ratio of H/D, which was also reported by Serrano (2015).

Figure 3b shows the radial velocity in dimensionless form at different positions of z/H. The predictive empirical model of the dimensionless radial velocity was developed by Beltaos and Rajaratnam (1974), which is expressed as Eq. (6) and shown in Fig. 3b as well. It is seen that the simulated results of the radial velocity match the values from the empirical model well.

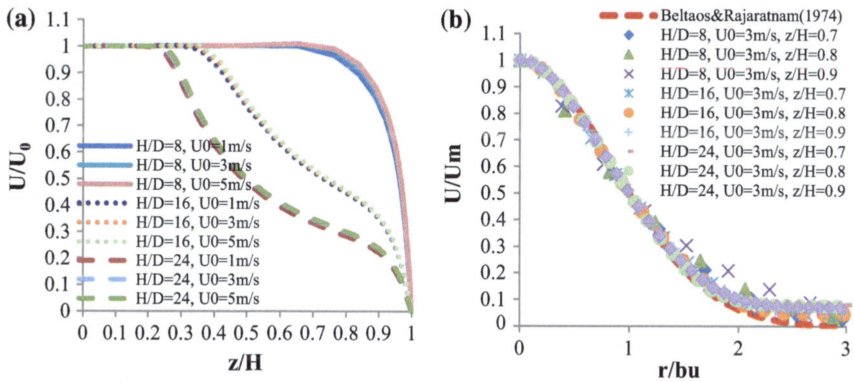

Fig. 3. Profiles of **a** centerline velocity and **b** radial velocity of the simulated jet flow

$$\frac{U}{U_m} = e^{-0.693(r/b_u)^2} \tag{6}$$

where U is the velocity along the radial direction; U_m is the maximum velocity at different height ratios of z/H; b_u is the distance value along radial direction where $U = 0.5U_m$.

The wall shear stress profile and the maximum wall shear stress values are considered the critical characteristics that can cause and influence the sand erosion under the impinging jet flow. The empirical model of the dimensionless wall shear stress distribution is expressed as (Beltaos and Rajaratnam 1974):

$$\frac{\tau_0}{\tau_{0m}} = 0.18\left(\frac{1 - e^{-114(r/H)^2}}{r/H}\right) - 9.43\left(\frac{r}{H}\right)e^{-114(r/H)^2} \tag{7}$$

where τ_0 is the wall shear stress along the radial direction; and τ_{0m} is the maximum shear stress and expressed as: $\tau_{0m} = 0.16\frac{\rho U_0^2}{\left(\frac{H}{D}\right)^2}$.

The wall shear stress and the dimensionless of $\frac{\tau_0}{\tau_{0m}}$ can also be expressed using a quadratic model as (Weidner et al. 2012):

$$\tau_0 = C_f \rho U_x^2 \tag{8}$$

$$\frac{\tau_0}{\tau_{0m}} = \frac{C_f \rho U_x^2}{C_f \rho U_{xmax}^2} = \frac{U_x^2}{U_{xmax}^2} \tag{9}$$

where C_f is the friction coefficient; ρ is the density of jet fluid; U_x is the near wall velocity parallel to the sand bed; and U_{xmax} is the maximum near wall velocity parallel to the sand bed along the radial direction. In this study, the near wall velocities of U_x and U_{xmax} are determined based on the CFD analysis at the location of z/H = 0.99, which is 0.01 H above the bottom wall.

The comparison between the dimensionless wall shear stress distribution determined from the empirical model and the CFD results is shown in Fig. 4a. The simulated results of all cases with different ratios of H/D and jet Reynolds numbers reasonably match the profile of the wall shear stress distribution determined from the empirical model. And the maximum shear stress occurs approximately at $r/H = 0.14$ for both empirical model and CFD simulation results. In addition, the absolute values of τ_{0m}/ρ derived from the empirical expression of τ_{0m} and the quadratic shear stress model are plotted in Fig. 4b with the fitting value of C_f equal to 0.01. It is seen that the calculated and simulated values of τ_{0m}/ρ follow the line of equity for all the CFD simulated cases at different ratios of H/D and jet Reynolds numbers. These results demonstrate that the developed CFD analysis can well characterize the key features of the impinging jet flow, in terms of centerline velocity, radial velocity, and wall shear stress distribution, under different ratios of H/D and jet Reynolds numbers. This analysis provides a prerequisite to implement a reliable coupled CFD and DEM analysis to simulate sand erosion under the impinging jets with the confidence of impinging jet flow behaviors.

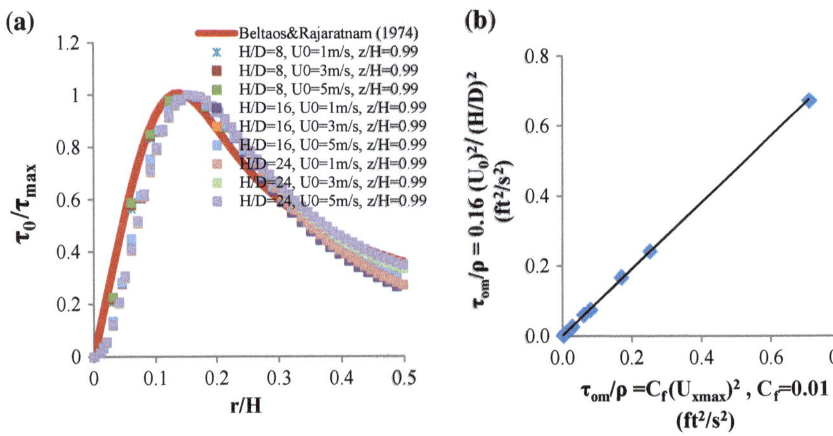

Fig. 4. Comparison of **a** wall shear stress distribution and **b** the absolute value of τ_{max}/ρ between the simulated results and the empirical models developed by Beltaos and Rajaratnam (1974)

3.2 Coupled CFD and DEM Analysis of Sand Erosion by an Impinging Jet

A snapshot of the coupled CFD and DEM analysis of the sand scour and the jet flow velocity vector at the middle cross-section plane is shown in Fig. 5, for the case of H/D = 16 and U_0 = 3 m/s at time of 15 s. The coupled CFD and DEM analysis can explicitly consider the interaction between the scour hole and jet flow, as the sand erosion is seen and a vortex forms near the scour hole. This vortex flow affected by the scour hole cannot be identified from the jet flow velocity field simulated via the individual CFD analysis as shown in Fig. 1b, in which the presence of soil particles and the effect of scour holes on jet flow were not considered.

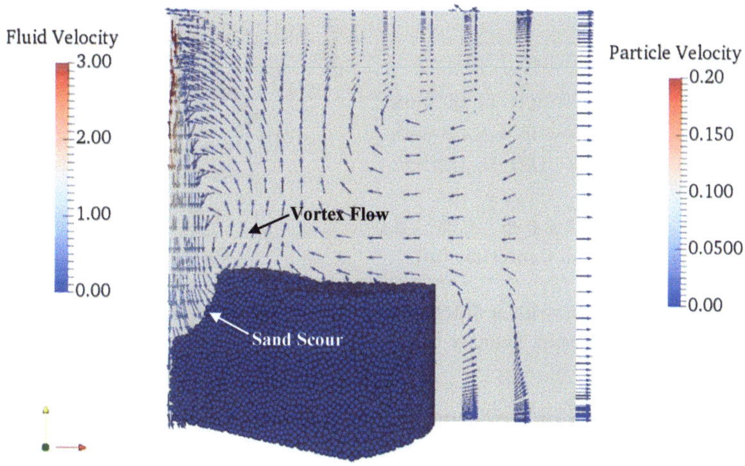

Fig. 5. A snapshot of sand erosion and jet flow velocity vector at 15 s for the case of H/D = 16 and U_0 = 3 m/s

The sand scour shapes at the middle cross-section plane for different nozzle inlet velocities at the ratio of H/D = 16 are shown in Fig. 6a. And Fig. 6b shows the sand scour shapes at different ratios of H/D but with the same nozzle inlet velocity of 3 m/s. The results show that as the nozzle inlet velocity increases, the maximum scour depth increases; and as the ratio of H/D increases, the maximum scour depth decreases. The results of the maximum scour depth, ridge height, and distance from centerline to ridge for different ratios of H/D and nozzle inlet velocities are presented in Table 2. The ridge of the scour hole, which is seen from the coupled CFD and DEM analysis, can also be observed in laboratory experiments (e.g., Aderibigbe and Rajaratnam 1996). This demonstrates that the developed coupled CFD and DEM analysis is a promising numerical approach to investigate the sand erosion behaviors under the impinging jet test, with the capability to capture the details of both jet flow behaviors and sand erosion characteristics.

Figure 6b also shows that at the ratio of H/D = 8 and U_0 = 3 m/s, the shape of the sand scour hole is narrow, deep, and localized. As the ratio of H/D increases, the shallow but wide scour holes are observed. This phenomenon has been reported in the experiments conducted by Moore and Masch (1962) that two different shapes of the scour hole were created depended on the H/D values, where H/D = 7 was a critical threshold value for the shape changes of the scour hole. Aderibigbe and Rajaratnam (1996) classified scour hole geometry as Strongly Deflected Jet Regime (SDJR) and Weakly Deflected Jet Regime (WDJR). Similar experimental observations were also seen in Mazurek et al. (2001) and Hou et al. (2016).

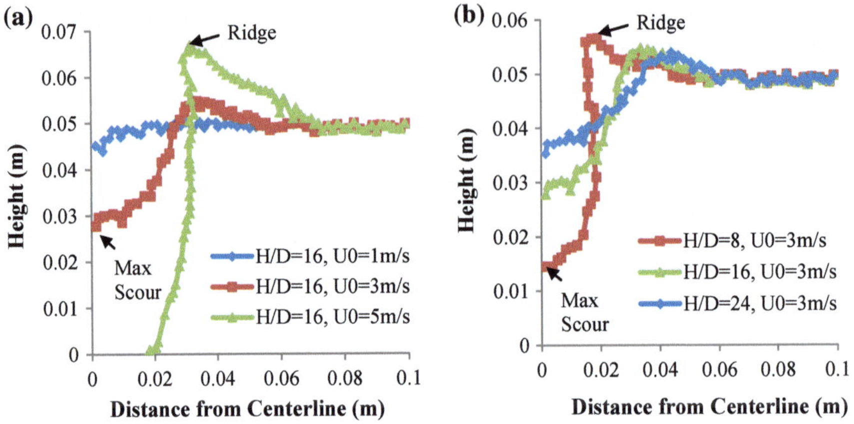

Fig. 6. Sand erosion profiles for **a** different nozzle inlet velocities at H/D = 16 and **b** different ratios of H/D with nozzle inlet velocity U_0 = 3 m/s

Table 2. Simulated results of maximum scour depth, ridge height, and distance from centerline to ridge for different ratios of H/D and nozzle inlet velocities

H/D	8			16			24		
U_0, m/s	1	3	5	1	3	5	1	3	5
Case number	1	2	3	4	5	6	7	8	9
Max. scour depth (ε_m), cm	0.855	3.550	NA (>5)	0.605	2.215	NA (>5)	0.199	1.464	3.522
Ridge height ($\Delta\varepsilon$), cm	0.097	0.660	0.734	0.027	0.486	1.690	0.027	0.398	1.094
Distance from centerline to ridge (x_c), cm	2.069	1.900	1.340	3.448	3.373	3.160	3.448	4.396	4.848

To investigate the mechanisms of SDJR and WDJR that generate different shapes of the scour hole, the simulation cases with ratios of H/D = 8 and H/D = 16 with U_0 = 3 m/s were selected to study the SDJR and WDJR in details based on the fluid-particle interaction forces from the coupled CFD and DEM analysis. The evolutions of the maximum fluid-particle interaction forces and the distributions of the fluid-particle interaction forces at the surface of sand bed for the two cases are plotted in Figs. 7 and 8. From the plots of the maximum fluid-particle interaction forces (see Figs. 7a and 8a), the scour process can be divided into three stages: scour onset, scour developing, and scour stabilization. At time of scour onset (see Figs. 7b and 8b), the distribution of the fluid-particle interaction forces along the surface of the sand bed show that the maximum interaction force occurs at 0.016 and 0.023 m from the nozzle axis of the jets for the ratios of H/D = 8 and 16, respectively. The influence range of the interaction forces becomes wider but the peak force value decreases as the H/D increases.

During the scour developing stage, the maximum interaction force fluctuates due to the kinematics of the sand particles. And the common feature of the distribution of

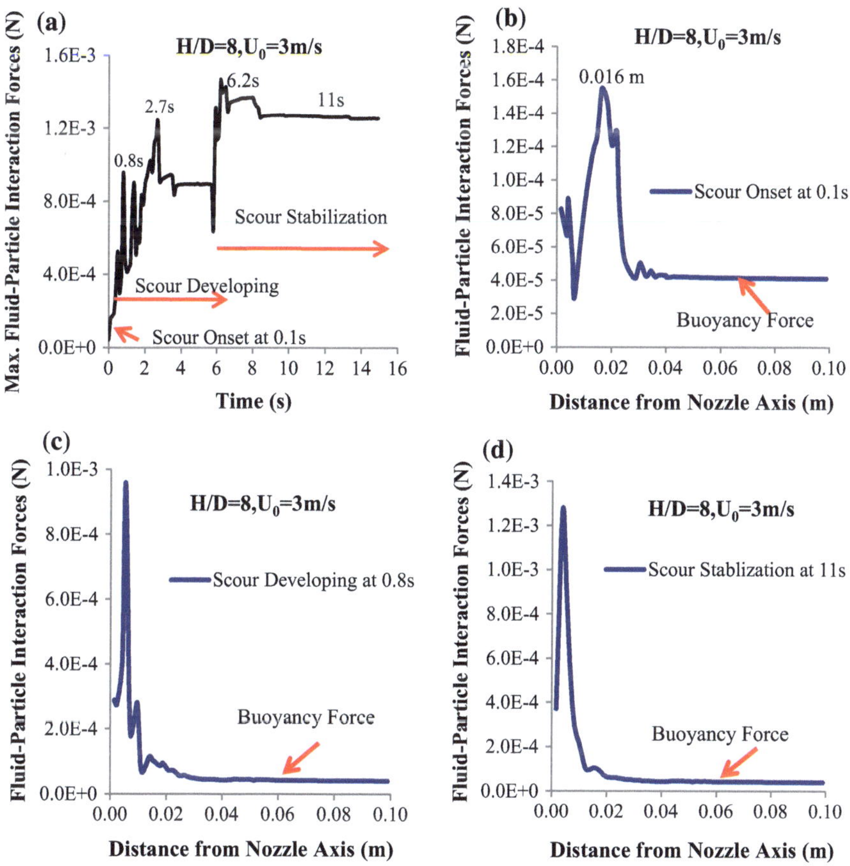

Fig. 7. Evolution of **a** the maximum fluid-particle interaction forces and the distribution of the fluid-particle interaction forces at **b** scour onset, **c** scour developing, and **d** scour stabilization for the case of H/D = 8 and U_0 = 3 m/s

interaction forces at this stage is that the peak interaction force occurs near the nozzle axis, as seen from Figs. 7c and 8c. As the interaction between jet flow dynamics and sand particles kinematics reaches an equilibrium status, the erosion process ceases. The scour shape tends to stabilize and the maximum interaction force reaches a constant value. The representative distributions of the interaction forces at the scour stabilization stage are plotted in Figs. 7d and 8d. For the impinging jet with ratio of H/D = 8, a single and higher peak force occurs near to the nozzle axis, when compared with the impinging jet with ratio of H/D = 16, where double peak forces and the wider range of influence are observed. In addition, the buoyancy force can be recognized specifically on the soil particles located far from the nozzle axis of the jet, which can be seen in Figs. 7 and 8.

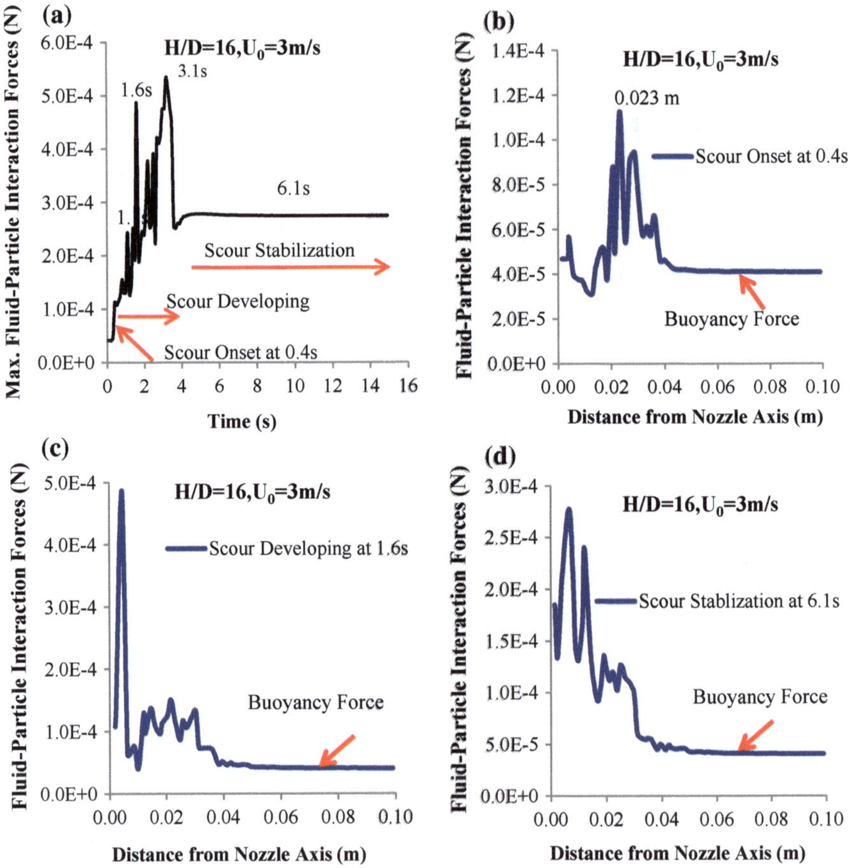

Fig. 8. Evolution of **a** the maximum fluid-particle interaction forces and the distribution of the fluid-particle interaction forces at **b** scour onset, **c** scour developing, and **d** scour stabilization for the case of H/D = 16 and U_0 = 3 m/s

In summary, the proposed coupled CFD and DEM analysis can explicitly incorporate the interaction between the jet flow and sand particles erosion, which is essential for the soil erosion analysis at the particle scale. The sand scour development process and the final scour shape under the impinging jet flow can be analyzed directly based on the coupled CFD and DEM analysis. As the ratio of H/D increases, the influence range of the interaction forces increases but the value of peak force decreases. This explains the mechanisms of the formations of different shapes of scour hole under the strongly and weakly deflected jets, which originally contributes to the understanding of the sand erosion behaviors under the impinging jet flow at the particulate scale.

4 Conclusions

In this study, the coupled CFD and DEM analysis was proposed and established based on the prerequisite of the individual CFD analysis to simulate the sand erosion process under the impinging jet flow. The fluid-particle interaction forces, including Di Felice drag force, pressure gradient force, viscous force, and buoyant force, are explicitly considered in the numerical analysis of the sand erosion process. The developed CFD analysis can well capture the key features of jet flow behaviors, such as the dimensionless profiles of centerline velocity, radial velocity, and the wall shear stress distribution. The simulation results of the coupled CFD and DEM analysis are as expected with the maximum scour depth increases as the ratio of H/D decreases and nozzle inlet velocity increases. The observations of different shapes of the scour hole from the numerical analysis follow the findings from the experiments related to the SDJR and WDJR. The explanations of the mechanisms of SDJR and WDJR were discussed based on the numerical analysis results on the fluid-particle interaction forces at the particulate scale, which were obtained from the coupled CFD and DEM analysis. This helps to understand the sand erosion mechanisms under the jet flow. Thus, the proposed coupled CFD and DEM analysis is a promising tool to analyze the fundamental erosion mechanisms of the onset and propagation in the soil erosion process. The limitations of this work include that the input DEM parameters of the sand particles are not specifically calibrated, and the original thickness of sand bed is shallow for several simulated cases so that the maximum scour depth cannot be captured. Future works also need to consider the bonding and debonding forces for the stabilized sand materials and cohesive clays.

References

Aderibigbe, O.O., Rajaratnam, N.: Erosion of loose beds by submerged circular impinging vertical turbulent jets. J. Hydraul. Res. **34**(1), 19–33 (1996). https://doi.org/10.1080/00221689609498762

Beltaos, S., Rajaratnam, N.: Impinging circular turbulent jets. J. Hydraulics Div. **100**(10), 1313–1328 (1974)

CFDEM®project: https://www.cfdem.com/ (2017). Accessed by 15 Aug

Chen, J., Wang, Y., Li, X., et al.: Erosion prediction of liquid-particle two-phase flow in pipeline elbows via CFD–DEM coupling method. Powder Technol. **282**, 25–31 (2015)

Generic EDEM Material Model (GEMM) Database: https://www.edemsimulation.com/gemm/ (2017). Accessed by 15 Aug

Hanson, G.J., Cook, K.R.: Apparatus, test procedures, and analytical methods to measure soil erodibility in situ. Appl. Eng. Agric. **20**(4), 455–462 (2004)

Hanson, G.J., Hunt, S.L.: Lessons learned using laboratory jet test method to measure soil erodibility of compacted soils. Appl. Eng. Agric. **23**(3), 305–312 (2006)

Hanson, G.J., Robinson, K.M., Temple, D.M.: Pressure and stress distributions due to a submerged impinging jet. Hydraul. Eng., 525–530 (1990)

Hou, J., Zhang, L., Gong, Y., et al.: Theoretical and experimental study of scour depth by submerged water jet. Adv. Mech. Eng. **8**(12), 1–9 (2016). https://doi.org/10.1177/1687814016682392

Kuang, S.B., LaMarche, C.Q., Curtis, J.S., Yu, A.B.: Discrete particle simulation of jet-induced cratering of a granular bed. Powder Technol. **239**, 319–336 (2013)

Mazurek, K.A., Rajaratnam, N., Sego, D.C.: Scour of cohesive soil by submerged circular turbulent impinging jets. J. Hydraul. Eng. **127**(7), 598–606 (2001)

Mercier, F., Bonelli, S., Anselmet, F., et al.: On the numerical modelling of the jet erosion test. In: The 6th International Conference on Scour and Erosion, Paris, France, pp. 601–608 (2012)

Mercier, F.: Numerical modelling of erosion of a cohesive soil by a turbulent flow. Doctoral dissertation, Aix-Marseille Université (2013)

Moore, W.L., Masch Jr., F.D.: Experiments on the scour resistance of cohesive sediments. J. Geophys. Res. **67**(4), 1437–1449 (1962)

OpenFOAM-3.0.1: User Guide. OpenFOAM Foundation Ltd. (2015)

Serrano, V.M.C.: Numerical Investigation of the Shear Stress Distribution Resulting from a Turbulent Impinging Jet. Master thesis, Washington State University (2015)

Weidner, K., Petrie, J., Diplas, P., et al.: Numerical simulation of jet test and associated soil erosion. In: The 6th International Conference on Scour and Erosion, Paris, France, pp. 609–616 (2012)

Weidner, K.L.: Evaluation of the jet test method for determining the erosional properties of cohesive soils: a numerical approach. Master thesis, Virginia Polytechnic Institute and State University (2012)

Zhang, Y., Zhao, M., Kwok, K.C.S., Liu, M.M.: Computational fluid dynamics-discrete element method analysis of the onset of scour around subsea pipelines. Appl. Math. Model. **39**, 7611–7619 (2015)

Zhao, J., Shan, T.: Coupled CFD-DEM simulation of fluid-particle interaction in geomechanics. Powder Technol. **239**, 248–258 (2013)

Zhou, Z.Y., Kuang, S.B., Chu, K.W., Yu, A.B.: Discrete particle simulation of particle-fluid flow: model formulations and their applicability. J. Fluid Mech. **661**, 482–510 (2010). https://doi.org/10.1017/S002211201000306X

Performance Studies of Microbial Induced Calcite Precipitation to Prevent the Erosion of Internally Unstable Granular Soils

Fatima-Zahra Haouzi[1(✉)], Annette Esnault-Filet[2], and Benoît Courcelles[3]

[1] Polytechnique Montréal, Montreal, QC, Canada
fatima-zahra.haouzi@polymtl.ca
[2] Soletanche-Bachy, Rueil-Malmaison, France
[3] CGM Department, Polytechnique Montréal, Montreal, QC, Canada

Abstract. Migration of fine particles within internally unstable granular soils under water seepage flow (suffusion) is one of the most common causes of earth infrastructures' failure. To assess the ability of Microbial Induced Calcite Precipitation (MICP) to prevent the segregation in an internally unstable soil, internal erosion tests were conducted upon soil samples treated by bacteria and cementation solutions. MICP experiments were carried out with concentrations of urea/$CaCl_2$ equal to 1.4 M. Volumes of injected bacteria solutions were equal to the volumetric water content corresponding to different tested degrees of saturation (S_r): 30, 60 and 80%. Cementation solutions were injected three times for each sample. Biochemical properties of MICP were examined to predict bacterial movement through soil matrices as a function of S_r. The amount of the $CaCO_3$ produced was examined depending on S_r. Following their treatment, samples were saturated and submitted to increment of hydraulic gradients varying from 0.1 to 10. Eroded fine particles masses, seepage flow rates and effective hydraulic gradients along samples were measured throughout the experiment. The results of our study pointed out that MICP stabilized internally unstable granular soils as the critical gradient went from 0.7 for untreated samples to 5 for biocemented samples.

1 Introduction

Microbially Induced Calcite Precipitation (MICP) is a biogrouting process that emerged in the last 20 years. This new soil improvement method is cost effective, easy to implement and environmentally friendly compared to traditional grouting techniques (DeJong et al. 2013; Ivanov and Chu 2008; Suer et al. 2009). The MICP technique is based on natural biochemical processes to improve engineering properties (strength and impermeability) of soils thanks to the precipitation of calcite on soil particle surfaces and at particle contacts. Calcite precipitation may be achieved by many different processes namely urea hydrolysis, denitrification, etc. Nevertheless, enzymatic hydrolysis of urea by microbes is the most energy efficient process and a wide variety of microorganisms can be used for their urease activity. The mechanism of $CaCO_3$ precipitation follows two stages: (1) urea hydrolysis into ammonium and carbonate

D. Chen et al. (eds.), *Advancements on Sustainable Civil Infrastructures*,
Sustainable Civil Infrastructures, https://doi.org/10.1007/978-3-319-96241-2_4

ions, and (2) formation of calcite in presence of calcium ions according to the chemical reactions below (DeJong et al. 2006):

$$CO(NH_2)_2 + H_2O \xrightarrow{urease} 2NH_4^+ + CO_3^{2-} \quad (1)$$

$$Ca^{2+} + CO_3^{2-} \rightarrow CaCO_3 \downarrow \quad (2)$$

The $CaCO_3$ crystallographic patterns, size, shape and distribution, directly impact the response of treated soils. Bacteria type, bacterial and cementation solutions, injection protocol, temperature, pH level, urease activity/availability of nucleation sites, degree of saturation and concentration of cementation solutions constitute factors that affect MICP performances (Mujah et al. 2016). Furthermore, homogeneity of calcite precipitation is linked to the injection protocol and the aeration of bacteria. Strength improvement of MICP treated sand is enhanced when exposed to the atmosphere (Amin et al. 2017; Cheng and Cord-Ruwisch 2012; Cheng et al. 2013; Palmen 2012). Also, higher strengths are obtained in unsaturated conditions (Cheng et al. 2013) and the treatment by surface percolation gives an homogeneous distribution of calcite precipitation (Cheng and Cord-Ruwisch 2012).

The precipitated calcium carbonates only create bridges but do not fill the pores entirely. The effect on the hydraulic conductivity is thus limited (Esnault-Filet et al. 2016). To evaluate the scalability of MICP for an implementation in situ, some researchers have performed fields trials and upscaled experiments (10–100 cm in length) (Van Paassen et al. 2009a), large scale columns (2.5 m length) and pilot scales (5 m spacing) (Esnault-Filet et al. 2016; Van Paassen 2009). In all cases, authors found out that the geotechnical properties of biocemented sand were highly enhanced. Biocementation spatial distribution homogeneity remains problematic though.

MICP was successfully applied for purposes from increasing stiffness and strength of sand (DeJong et al. 2010; Filet et al. 2012; Van Paassen et al. 2010), enhancing slope stability and bearing capacity (DeJong et al. 2013), reducing foundation settlements, increasing pore pressure to the mitigation of liquefaction (Esnault-Filet et al. 2016; Montoya and DeJong2015) and preventing soil from erosion (Amin et al. 2017; Ning-Jun et al. 2017). As regards this last point, the action of water can be particularly harmful for roads and airfields (Chapuis et al. 1996). Seepage-induced erosion involves a selective migration of fines particles through the coarse skeleton of a granular soil (suffusion) and the accumulation of these particles at the outlet may diminish the hydraulic performances. As consequence, problems linked to frost, settlement, cracking may occur. MICP has the potential to prevent fines migration in a non-intrusive way maintaining a relative high permeability. Ning-Jun et al. (2017) and Amin et al. (2017) proved that biomineralization mitigates the hydraulic erosion of sand. Thus, the higher the precipitation content is, the larger the size of clusters of cemented sand particles can be obtained, preventing sand particles from being flushed out.

In the present work, the mitigation of erosion through MICP is assessed in an internally unstable sand at different degrees of saturation. Biotreated samples were subjected to increments of hydraulic head using a custom erosion test procedure. The aim of this investigation is to evaluate the applicability of MICP to roads and pavements.

2 Materials and Methods

2.1 Tested Materials

2.1.1 Tested Soil

The tested material was made of a mixture of granular materials of different particle sizes (70% of gravel and 30% of fine sand). The grain size distribution curve obtained fits within the gradation limits required by the Quebec Ministry of Transportation (MTQ) in Canada for subbase aggregates (Fig. 1). The resulting material is classified SP (silty sand) according to U.S.C.S. Its coefficient of uniformity C_u is equal to 13 which makes it poorly graded soil with concave upward gradation curve [internally unstable according to Lafleur et al. (1989) and Chapuis et al. (1996)].

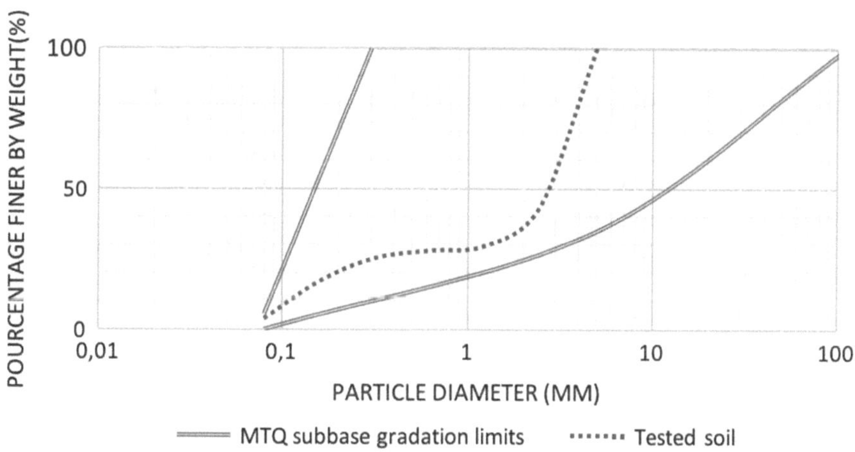

Fig. 1. Tested soil gradation curve

2.1.2 Bacterial Solutions

The calcifying bacteria used in this study was provided by Soletanche-Bachy. The company developed its own industrial method of biomineralization process that consists of the use of lyophilisat of *S. pasteurii* easy to implement at large scales. The industrial process is so-called Biocalcis (Esnault-Filet 2016). Note that *S. pasteurii* is an alkalophilic bacterium with a highly active urease enzyme and a high resistance to physical and chemical agents which makes it usable in situ. The bacterial solution's optical density (OD_{600}) was measured by a spectrophotometer at 600 nm wavelength. The OD_{600} values vary between 1.9 and 2.1. The measured specific urease activity was ranging from 370 to 1100 (mS/cm/min/OD).

2.1.3 Fixation and Cementation Solutions

An equimolar urea-calcium chloride cementation solution was prepared at concentration of 1.4 M. A fixation solution was prepared of equimolar urea-calcium chloride concentration of 0.35 M.

2.2 Experimental Procedure

2.2.1 Preparation of Sand Specimen

Plexiglass moulds of 100 mm inner diameter and 200 mm length were used to prepare 160 mm high sand test specimens. After preparing the dry mixture of the tested soil, the material was wet with the equivalent volume of tested saturation (S_r) of water specifically 30, 60 and 80% (Table 1). The partially saturated soil was left to stand for at least 2 h. The material was then deposited in the mould in 5 layers without any compaction to prevent any segregation.

2.2.2 Biotreatment Process

In all experiments, the bacterial, fixation and cementation solutions were injected into the sand specimens by percolation from the surface. The reagents were introduced from the top of the columns and the transport of liquids was the result of gravity and capillary forces. The percolation method consisted of the following steps (Table 1):

1. Percolation of a volume of bacterial suspension equivalent to the volume of water in the specimen ($V_{inj} = S_r * V_p$) at 25 mL/min. Effluent samples were collected to trace the bacteria leaching and OD_{600} was measured every 2 min.
2. After an hour, percolation of a volume of fixation solution equivalent to the volume of water initially in the specimen ($V_{inj} = S_r * V_p$). Effluent was sampled to trace the bacteria leaching and OD_{600} was measured every 2 min.
3. After 24 h, percolation of one pore volume (V_p) of cementation solution at a flow rate of 25 mL/min, equivalent to a mean flow velocity of 0.32 m/h. This step was repeated three times after 24, 48 and 72 h.

At the end of each step the bottom valve was closed to allow reactions to have place within specimens.

2.2.3 Erosion Test

An experimental set-up was customised to characterize the internal erosion of bio-treated sand samples (Fig. 2). Before applying hydraulic gradient, the saturation of samples was achieved by the application of a partial vacuum and the introduction of cleared water from the bottom to avoid any stress of the treated samples. A downward flow was then applied to the saturated samples with hydraulic gradient increased by increments from 0.1 to 10. Each hydraulic gradient was applied for 10 min. Eroded fine particles masses leached out of samples, seepage flow rates and effective hydraulic gradients along samples were measured throughout the experiment. To compare hydraulic parameters before and after erosion, hydraulic gradients were decreased gradually until there was no difference in hydraulic head between the entry and the exit of the samples.

2.2.4 Calcite Distribution

To investigate the amount and spatial distribution of the calcite precipitation along specimens, an hydrochloric acid dissolution technique was used. Following each erosion tests, sub-samples were collected at 2, 4, 8, 12, and 14 cm from the top of bio-treated specimens and oven-dried at 100 °C for 24 h. The dry mass (M_{sb}) was then

Table 1. Details of bio-treatment procedure for each sample

Sample reference	Initial saturation S_r (%)	Bacterial solution (SB)		Fixation solution (SF)		Cementation solution (SC)	
3014	30	$V_{inj}^{a} = 0.3 * V_p$ $OD_{600} = 2.0 \pm 0.1$	Batch time + 1 h.	$V_{inj} = 0.3 * V_p^{b}$ [urea/$CaCl_2$] = 0.35 M	Batch time = 24 h	$V_{inj} = 1 * V_p$ [urea/$CaCl_2$] = 1.4 M This step is repeated 3 times	Batch time = 24 h
6014	60	$V_{inj} = 0.6 * V_p$ $OD_{600} = 2.0 \pm 0.1$		$V_{inj} = 0.6 * V_p$ [urea/$CaCl_2$] = 0.35 M			
8014	80	$V_{inj} = 0.8 * V_p$ $OD_{600} = 2.0 \pm 0.1$		$V_{inj} = 0.8 * V_p$ [urea/$CaCl_2$] = 0.35 M			

[a]V_{inj} is volume of the injected solution
[b]V_p is the specimen pore volume

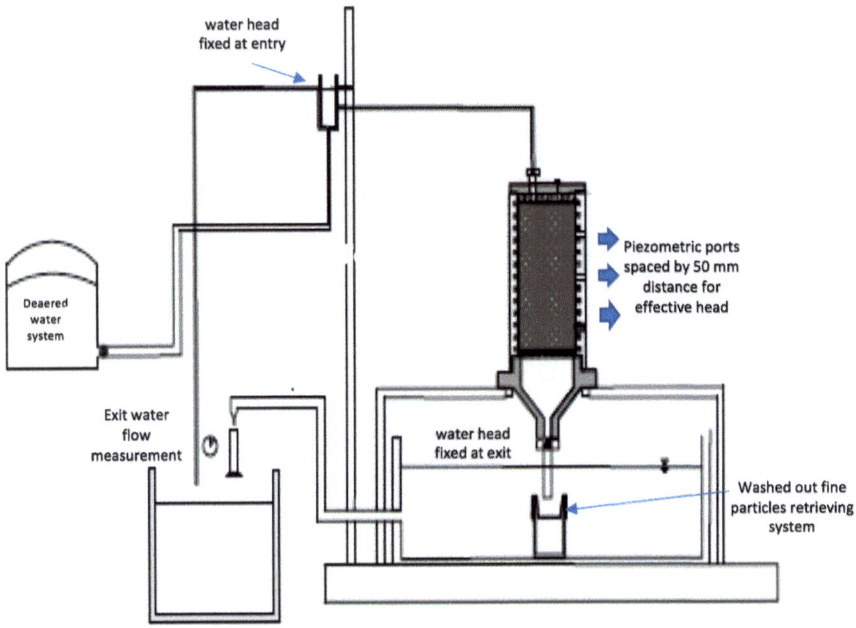

Fig. 2. Erosion test experimental set-up

weighed before being soaked in a 3-pH hydrochloric solution for 24 h. The effluent was drained using a filter paper and the remaining material was oven-dried for 24 h before being weighed (M_{sa}) to determine the calcium carbonate content following Eq. (3).

$$CaCO_3\ \text{content}(\%) = \frac{M_{sb} - M_{sa}}{M_{sa}} \times 100\% \qquad (3)$$

3 Results and Discussion

3.1 Bacteria Leaching

The measurement of the OD_{600} at the effluent during the injection test revealed the concentrations of bacteria leached out of the specimens. Figure 3 illustrates the normalized bacteria concentrations measured as a function of the number of pore volumes introduced in the samples for the different saturation degrees tested. The normalized bacteria concentration is calculated as the ratio of the measured OD_{600} on the initial OD_{600} of the injected solution. Bacteria started being leached when the equivalent of one pore volume was injected for all specimen tested (Fig. 3b). The concentration of leached bacteria dropped to less than 10% after 1.75 pore volume injected. This phenomenon didn't occur for treated samples at S_r = 30% as the total volume injected doesn't exceed one pore volume (Fig. 3a).

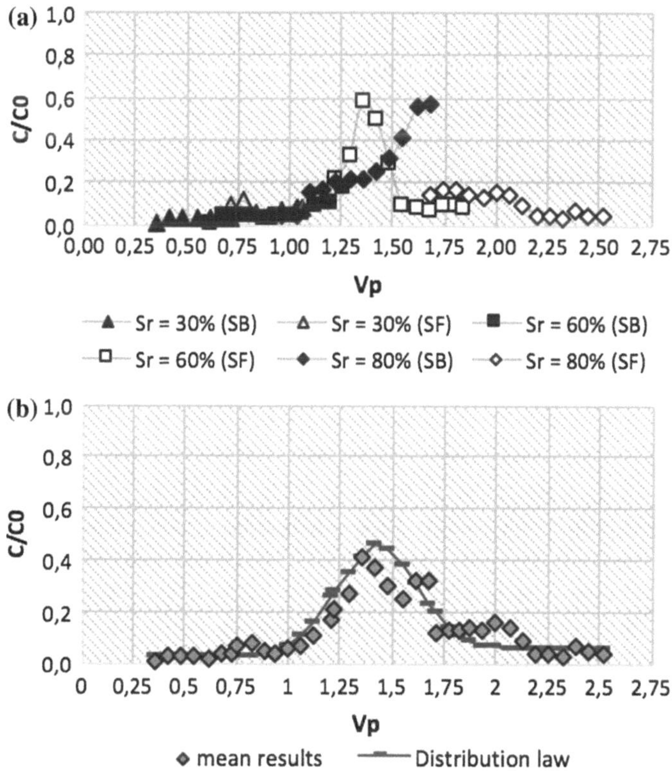

Fig. 3. Concentration of bacteria leached out of specimens during injection: **a** results for each specimen **b** mean values

3.2 Precipitated Calcite

The calcite content obtained at depths 2, 4, 8, 12 and 14 cm from the top are presented in Fig. 4. The amount of calcite depends on the volume of bacteria injected that's to say initial degree of saturation and the spatial distribution of calcite content is more uniform for biotreatment at 30% saturation. Also, the calcium content increases with the volume of bacteria and reaches 10% for the biotreatment at $S_r = 80\%$. In all cases, the amount of calcite is higher at the bottom of each specimen as a result of the percolation in the specimens.

3.3 Hydraulic Behavior of Treated Samples

3.3.1 Total Erosion Mass and Critical Hydraulic Gradients

The amount of fine particles washed out of samples was measured as a function of the applied hydraulic gradient (Fig. 5). The total mass (M_p) was normalized per unit area. The amount of washed out fine particles dropped significantly with MICP treatment. This parameter is directly linked to the volume of bacteria injected and the calcite content in the specimen. The total erosion mass is up to 36 times less than the untreated

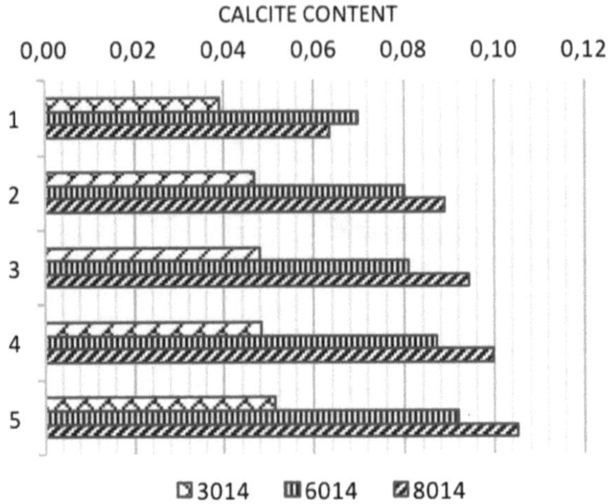

Fig. 4. Spacial distribution of calcite content in bio-treated specimens measured at (1) 2 cm; (2) 4 cm; (3) 8 cm; (4) 12 cm and (5) 14 cm from the top

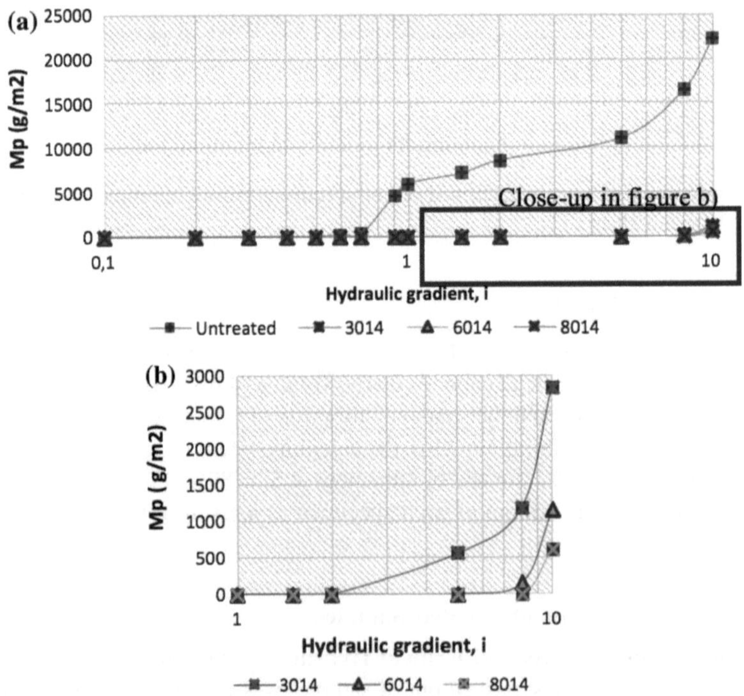

Fig. 5. Fine particles migration as function of applied hydraulic gradient; **a** all specimens, **b** close-up 3014, 6014, and 8014 specimens

Table 2. Erosion test results for all specimens

Sample reference	S_r (%)	i_{cr}	M_p (g/m^2)	$CaCO_3$ (%)
Untreated	–	0.7	22,330	–
3014	30	5	2838	4.7
6014	60	8	1151	8.2
8014	80	10	618	9.0

sample for a critical hydraulic gradient equal to 10 versus 0.7 for the untreated sand sample. The treatment at 30% gives already a fair response. The total mass eroded out of sample during the erosion test dropped to 2838 versus 22,330 g/m^2 for untreated samples. The theory of Lafleur et al. (1989) suggests that the 6014 and 8014 samples are at equilibrium (M_p is less than 2500 g/m^2).

The critical hydraulic gradient (i_{cr}) is defined as the gradient corresponding to initiation of suffusion that's to say the gradient when the fine particles are detected at the effluent. MICP treatment triggered a significant increase in the critical gradient as it is equal to 0.7 for untreated specimen while equal to 5 for S_r = 30% biotreatment. The 8014 specimen resists to a gradient equal to 8 and fine particles were detected at the outlet at a gradient of 10 (Table 2).

3.3.2 Seepage Water Flow and Hydraulic Conductivity

Water flow was measured during charging and discharging to assess the hydraulic behavior of sand specimens. Figure 6 illustrates the hydraulic behavior of untreated specimens, and 3014, 6014 and 8014 biotreated specimens respectively. For the untreated specimens, a clogging occurs at gradient of 0.4 and continues until suffusion happens at a gradient of 0.7, then the water flow increases from 0.1 to 3.97 cm^3/s. During discharging, the water flow measured under a gradient of 0.3 was equal to 7.98 cm^3/s compared to 0.1 cm^3/s during charging. This observation is due to the increase that occurs in pore volumes as a large amount of fine particles was leached out of the specimens. By contrast, bio-treated specimens showed an "elastic" behavior as no difference was revealed between seepage water flow during charging and discharging. The overall porosity of bio-treated specimens is unchanged following erosion which is comparable to results reported by Ning-Jun et al. (1610).

The measurement of hydraulic charge during erosion tests using the piezometric ports installed along specimens leads to the calculation of effective hydraulic conductivities of tested samples. Figure 7 shows the mean ratio between hydraulic conductivities at the top and the bottom of specimens during erosion tests. In general, treated samples showed a stable hydraulic conductivity between charging and discharging at top and bottom of specimen respectively. On the contrary, the hydraulic conductivity of untreated sand is much higher after the erosion of fine particles.

Table 3 shows the ratio between hydraulic conductivity measured at seepage charging (k_i) and seepage discharging (k_f) at an applied gradient equal to 0.3 before and after erosion. All the treated specimen show a fair stable hydraulic behavior whereas this ratio reaches 67 for untreated sand. This phenomenon is due to the increase of pore volume in the soil skeleton following the erosion of fine particles out of specimen

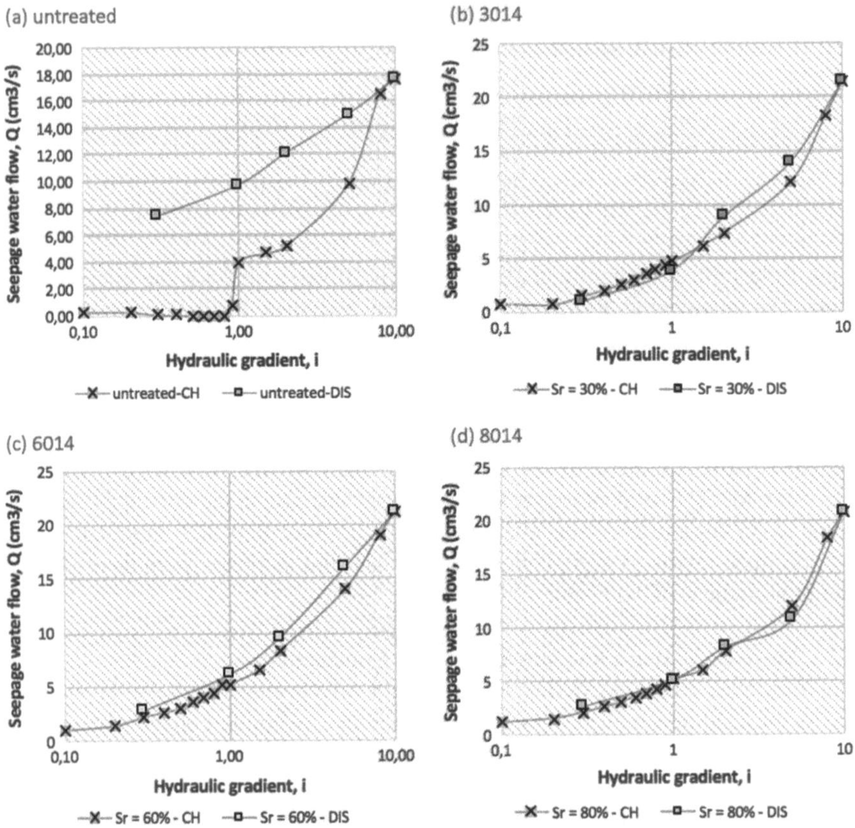

Fig. 6. Seepage water flow as a function of applied hydraulic gradient for **a** untreated sand, **b** 3014, **c** 6014 and **d** 8014 treated sand specimens

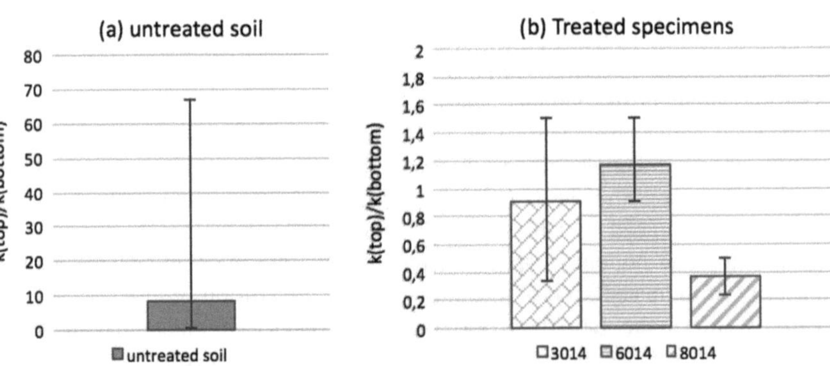

Fig. 7. Ratios of hydraulic conductivities between the top and bottom of specimens for the overall erosion test for **a** untreated sand and **b** bio-treated specimens

Table 3. Ratios of hydraulic conductivities before and after suffusion at hydraulic gradient equal to 0.3

	Untreated	3014	6014	8014
k_f/k_i (top)	67.6	1.4	2.1	1.1
k_f/k_i (bottom)	33.3	1.4	1.8	1.0

which does not occur for bio-treated sand samples. Cavity is increasing which weaken the soil leading to loosing strength and differential settlements. This could be explained as the effective stresses are decreased and pore pressure is increasing leading to fine particles migration and leaving cavities in the soil at a gradient lower than Terzaghi's hydraulic critical gradient (Skempton and Brogan 1994).

4 Conclusion

The experimental procedure followed herein reproduces the hydraulic response of a biocemented subbase material which was initially identified as internally unstable. The results of this investigation suggest that Microbially Induced Calcite Precipitation can significantly enhance the overall hydraulic behavior of a subbase material in terms of hydraulic critical gradient, outlet waterflow and amount of eroded fine particles.

Bio-treatment under unsaturated conditions gives a uniform spatial distribution of calcite along specimens and the migration of fine particles is prevented after only four days of treatment. The critical hydraulic gradient is improved as bio-treated specimens resist to a hydraulic gradient of, at least, 5 compared to 0.7 for untreated samples. Moreover, the volume of the injected bacterial solution was varied as a function of the initial saturation degree of each specimen. The results confirmed that the increase of the bacterial solution leads to higher calcite concentrations. However, higher concentrations of by-products (ammonium) are also released into environment. Therefore, the treatment at a degree of saturation of 30% is advantageous as it already shows a stable hydraulic behavior with a homogeneous distribution of calcite and minimizes the by-products.

Thus, an internally unstable soil can be valorized and used into soil foundations. In large scale applications, parameters such as volumes and concentrations of reagents, infiltration rate, number of injections and treatment duration should be customized depending on field conditions.

As a result, the treatment procedure is efficient in unsaturated conditions and can be applied by simple surface spraying in several infrastructures such as forest roads, culverts and wastewater pipes. MICP offers a huge benefit in cutting down roads' rehabilitation and maintenance costs.

Acknowledgements. This research was supported by the Natural Science and Engineering Research Council of Canada, the Fonds de Recherche de Québec—Nature et technologie and Menard Canada. This support is gratefully acknowledged. In addition, authors thank the support of Biocalcis® team for their involvement in the advancement of this investigation and the supply of the lyophilized bacteria.

References

Amin, M., Zomorodian, S.M.A., Okelly, B.C.: Reducing the hydraulic erosion of sand using microbial-induced carbonate precipitation. Proc. Inst. Civ. Eng. Ground Improv. **170**(2), 112–122 (2017). https://doi.org/10.1680/jgrim.16.00028

Chapuis, R.P., Contant, A., Baass, K.A.: Migration of fines in 0-20 mm crushed base during placement, compaction, and seepage under laboratory conditions. Can. Geotech. J. **33**(1), 168–176 (1996)

Cheng, L., Cord-Ruwisch, R.: In situ soil cementation with ureolytic bacteria by surface percolation. Ecol. Eng. **42**, 64–72 (2012). https://doi.org/10.1016/j.ecoleng.2012.01.013

Cheng, L., Cord-Ruwisch, R., Shahin, M.A.: Cementation of sand soil by microbially induced calcite precipitation at various degrees of saturation. Can. Geotech. J. **50**(1), 81–90 (2013). https://doi.org/10.1139/cgj-2012-0023

DeJong, J.T., Fritzges, M.B., Nusslein, K.: Microbially induced cementation to control sand response to undrained shear. J. Geotech. Geoenviron. Eng. **132**(11), 1381–1392 (2006). https://doi.org/10.1061/(asce)1090-0241(2006)132:11(1381)

DeJong, J.T., Mortensen, B.M., Martinez, B.C., Nelson, D.C.: Bio-mediated soil improvement. Ecol. Eng. **36**(2), 197–210 (2010). https://doi.org/10.1016/j.ecoleng.2008.12.029

DeJong, J.T., Soga, K., Kavazanjian, E., Burns, S., Van Paassen, L.A., Al Qabany, A., Weaver, T.: Biogeochemical processes and geotechnical applications: progress, opportunities and challenges. Geotechnique **63**(4), 287–301 (2013). https://doi.org/10.1680/geot.SIP13.P.017

Esnault-Filet, A., Gutjahr, I., Mosser, J.F., Sapin, L., Ibrahim, K.: A novel grouting process for the reinforcement of low permeability soils with the use of biocementation by biocalcis. Paper presented at the 19th Southeast Asian geotechnical Conference & 2nd AGSSEA Conference, Kuala Lumpur (2016)

Filet, A.E., Gadret, J.-P., Loygue, M., Borel, S.: Biocalcis and its applications for the consolidation of sands. Paper presented at the Geotechnical Special Publication, New Orleans, LA, United States (2012)

Ivanov, V., Chu, J.: Applications of microorganisms to geotechnical engineering for bioclogging and biocimentation of soil in situ. Rev. Environ. Sci. Biotechnol. **7**(2), 14 (2008)

Lafleur, J., Mlynarek, J., Rollin, A.L.: Filtration of broadly graded cohesionless soils. J. Geotech. Eng. Asce **115**(12), 1747–1768 (1989)

Montoya, B.M., DeJong, J.T.: Stress-strain behavior of sands cemented by microbially induced calcite precipitation. J. Geotech. Geoenviron. Eng. **141**(6), 04015019 (04015010) (2015). https://doi.org/10.1061/(asce)gt.1943-5606.0001302

Mujah, D., Shahin, M.A., Cheng, L.: State-of-the-art review of bio-cementation by microbially induced calcite precipitation (MICP) for soil stabilization. Geomicrobiol. J. **14** (2016)

Ning-Jun, J., Soga, K., Kuo, M.: Microbially induced carbonate precipitation for seepage-induced internal erosion control in sand-clay mixtures. J. Geotech. Geoenviron. Eng. **143**(3), 04016100 (04016114) (2017). https://doi.org/10.1061/(asce)gt.1943-5606.0001559

Palmen, A.: Stabilization of frictional soil through injection using CIPS (Calcite In-situ Precipitation System). (PhD), KTH Royal Institute of Technology, Stockholm, Sweden (2012)

Skempton, A.W., Brogan, J.M.: Experiments on piping in sandy gravels. Geotechnique **44**(3), 449–460 (1994)

Suer, P., Hallberg, N., Carlsson, C., Bendz, D., Holm, G.: Biogrouting compared to jet grouting: Environmental (LCA) and economical assessment. J. Environ. Sci. Health Part A Toxic/Hazard. Subst. Environ. Eng. **44**(4), 346–353 (2009). https://doi.org/10.1080/10934520802659679

Van Paassen, L.A.: Biogrout: ground improvement by microbially induced carbonate precipitation. (PhD), Delft University of Technology (2009)

Van Paassen, L.A., Ghose, R., van der Linden, T.J.M., van der Star, W.R.L., van Loosdrecht, M.C.M.: Quantifying biomediated ground improvement by ureolysis: large-scale biogrout experiment. J. Geotech. Geoenviron. Eng. **136**(12), 1721–1728 (2010). https://doi.org/10.1061/(asce)gt.1943-5606.0000382

Van Paassen, L.A., Harkes, M.P., Van Zwieten, G.A., Van Der Zon, W.H., Van Der Star, W.R.L., Van Loosdrecht, M.C.M.: Scale up of BioGrout: A biological ground reinforcement method. Paper presented at the Proceedings of the 17th International Conference on Soil Mechanics and Geotechnical Engineering: The Academia and Practice of Geotechnical Engineering, Alexandria, Egypt (2009a)

Van Paassen, L.A., Pieron, M., Mulder, A., van der Linden, T.J.M., Van Loosdrecht, M.C.M., Ngan-Tillard, D.J.M.: Strength and deformation of biologically cemented sandstone. Paper presented at the Rock engineering in difficult ground conditions-Soft rocks and karst, Dobrovnik, Croatia (2009b)

On the Determination of Undrained Shear Strength from Vane Test

Mounir Bouassida(✉) and Dalel Azaiez

Laboratoire de Recherche Lngénierie Géotechnique, Université de Tunis El Manar/Ecole Nationale D'lngénieurs de Tunis, LRl4ES03.BP 37 Le Belvédère, 1002 Tunis, Tunisia

mounir.bouassida@fulbrightmail.org, dalelazaiez@gmail.com

Abstract. Due to their low consistency and colloidal aspect, there have been many methods to determine the undrained shear strength c_u of soft clays. Aiming at meaningful method for c_u determination, a new method is suggested. In this work, focus is given on the vane shear test; one of the in situ test especially developed for preserving the soil structure and avoiding its disturbance to proceed for the determination of c_u. Precisely, the purpose was to discuss the credibility of the usually used method to calculate the undrained shear strength from in situ vane shear test and to confirm Bouassida proposed method (Proceedings of the 12th African Regional Conference on Soil Mechanics and Geotechnical Engineering. Durban, South Africa, pp. 285–291, 1999). Data of in situ vane shear tests conducted on Gothenburg, Tunis and Le Havre soft clays were compiled to compare between cu estimations by different methods of determination. Furthermore, the determination of c_u from existing correlations was considered, to discuss the suitable value of undrained shear strength of soft clays.

1 Introduction

According to the norms already established, the undrained shear strength is determined from the maximum value of the measured torque corrected for apparatus rod friction without any restriction on the corresponding angle of rotation.

As confirmed by Bouassida and Boussetta (1999), since for certain depth the maximum pick will appear beyond small strain due to the gained resistance under the soil proper weight with depth, mobilized soil capacity will be overestimated at the time of yielding.

Further, based on several yield analyses of Backfill, Cotest and Sanglerat (1983), had concluded that the value of undrained shear strength determined from the vane apparatus is overestimated (Bouassida and Boussetta 2000).

Wroth (1984) has confirmed that during vane shear test the mode of failure is comparable to that of direct simple shear, yet not a conventional one since the specimen is dissimilarly oriented; which also was confirmed by Mahmoud as stated Chandler (1988) (Fig. 1).

D. Chen et al. (eds.), *Advancements on Sustainable Civil Infrastructures*, Sustainable Civil Infrastructures, https://doi.org/10.1007/978-3-319-96241-2_5

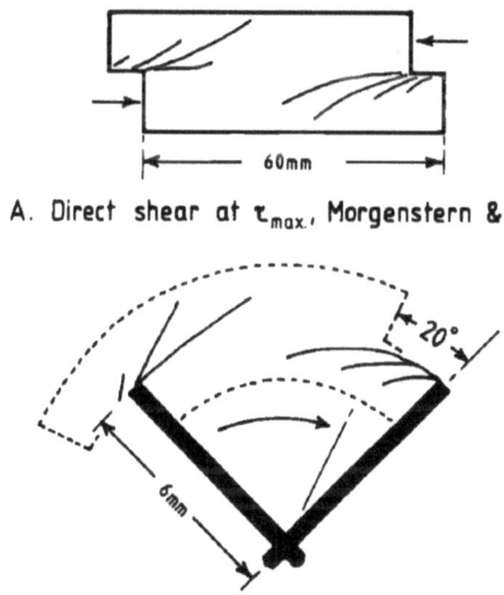

Fig. 1. Comparison of shear structures in direct shear test and vane shear test at maximum stress

In this frame Bouassida (2006) has proposed to limit the angle of rotation of the vane in order to obtain a coherent value of the measured torque. Hence, the shear strength will be determined as for the direct shear test with limitation of horizontal displacement.

Particularly, it consists of restricting the vane rotation angle α in a way that satisfies the following condition (Figs. 1 and 2)

$$\left\{\frac{\alpha_{max}}{90} = 6\% \text{ for } H = 2D \rightarrow \alpha_{max} = 5.4^{\circ}\right.$$

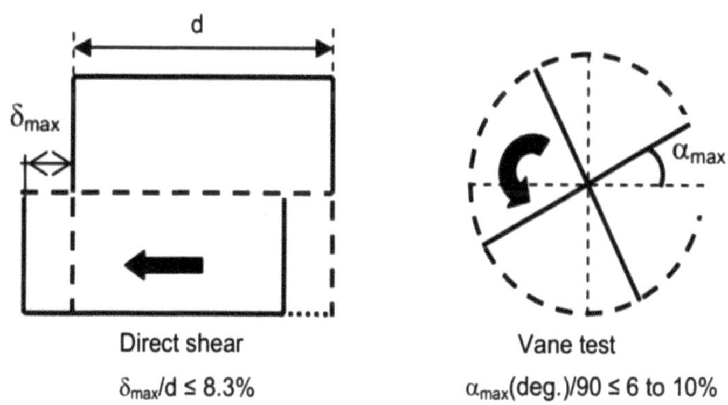

Fig. 2. Similarity between direct shear test and vane test Bouassida (2006)

In this frame, data of s_u measurements versus rotation angle taken with the vane shear test of Gothenburg clay (Sweden), Tunis soft clay (Tunisia), and Le Havre clay (France) were gathered and all available data required to characterize these sites.

The first section was dedicated to introduce these sites. The second section was to compare the obtained c_u measurements either from the furnished data of diverse test, particularly the in situ vane test based on the conventional method (maximal torque), and the torque corresponding to 20° and most importantly corresponding to Bouassida's proposed angle of rotation or through Skempton correlation (Eq. 1) when affordable;

$$\left(\frac{S_u}{\sigma'_{v0}}\right)_{NC} = 0.0037 I_p + 0.11 \quad (1)$$

The obtained results were interpreted by considering two parameters; the rotation angle and $\tan \lambda c_u$ in the third and fourth sections respectively.

It is important to mention that for Gothenburg clay, considering the depth of the investigation some vane curves measurement showed a secondary pick as illustrated in Fig. 3 (first plateau) which informs about the registered rod friction.

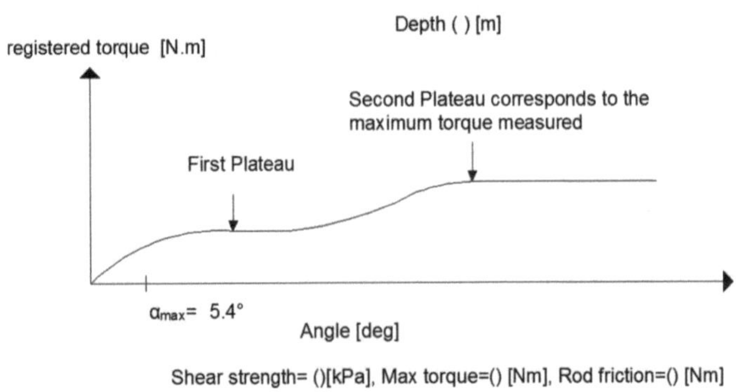

Fig. 3. Norconsult data annotations

Consequently, the undrained shear strength and the rotation angle corresponding to it (see Appendix) were calculated by removing rod friction effects and according to Eq. (2).

$$S_u = \frac{12T_{()}}{\pi D^2 \left(\frac{D}{\cos i_t} + \frac{D}{\cos i_b} + 6H\right)} \quad (2)$$

It is also noted that the vane is tapered lower end 11.0 × 5.0 cm; which means D = 5 cm, H = 11 cm, $i_t = 0\ rad(top\ angle\ of\ the\ vane)$; $i_b = \frac{\pi}{4} rad(bottom\ angle\ of\ the\ vane)$.

2 Sites Description

As mentioned earlier three different sites were investigated;

The Karlatower project (Karlavagnstornet) located in the Northern part of Lindholmen.

The investigated site is in Sweden. It is situated at the northern part of Lindholmen, Gothenburg in the vicinity of Lundbyleden (Fig. 4).

Fig. 4. Gothenburg site location

The soil is composed of 2 m padding materials at the surface followed with mighty clay layer.

Gothenburg clay is characterized with 60–75% moisture content. Its plasticity index varies between 30 and 45% and a unit weight of 15.94–16.94 $\frac{\text{kN}}{\text{m}^3}$. Its liquid limit varies 56–74% as illustrated in Fig. 5. Pre consolidation stress and initial stress as shown in Fig. 6 resulted in an OCR varying from 1.2 to 1.35 as shown in Fig. 7.

Using the available data, we concluded that the subjected clay is a soft soil which consistency index I_c does not exceed 0.4.

Tunis soft clay data mainly belongs to the exchanger project Cyrus le grand.

The investigated Tunisian site is a road exchanger project at Cyrus le Grand Avenue, located at the middle of Tunis at the extreme western part of the lace Le Lac (Fig. 8).

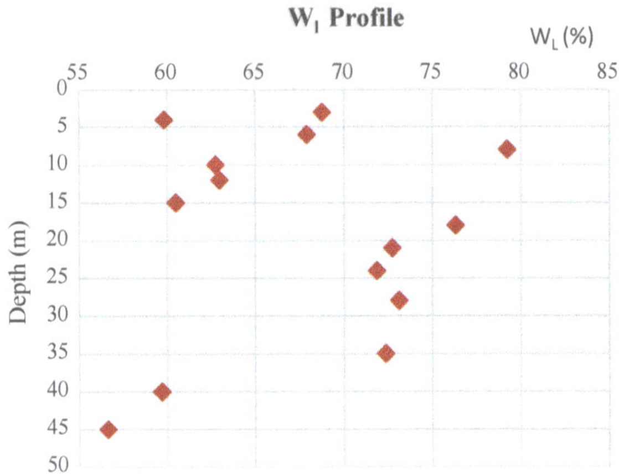

Fig. 5. Gothenburg clay W_l profile

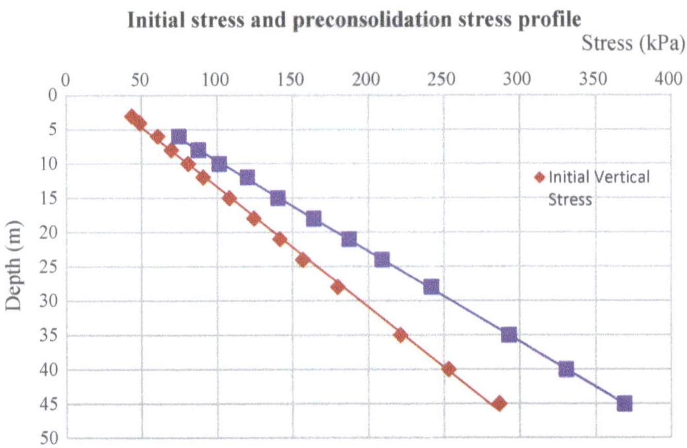

Fig. 6. Gothenburg soft clay preconsolidation and initial stress profile

Physical soil parameters investigation reached a depth of 45 m; It is composed of soft clay which unit weight γ_h varies from 1.42 to 1.95 $\frac{kN}{m^3}$, index plasticity varies from 17 to 35%, and its liquid limit varies from 14 to 70%.

French soft clay Vane shear test measurement were done in the vicinity of the harbor dock Le Havre in the estuary of the Seine (Fig. 9).

Considering the lack of data and in order to characterize the investigated site, reference was made to Troalen and all work (1996) which subjected the Seine's estuary's clay, precisely Le Havre harbor.

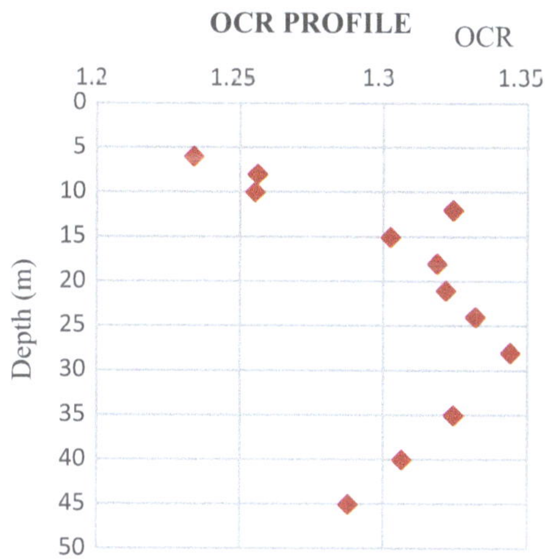

Fig. 7. Gothenburg soft clay OCR profile

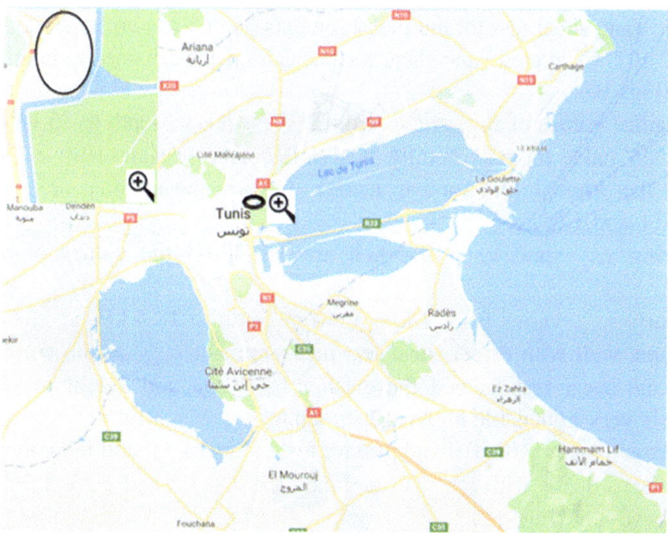

Fig. 8. Tunis site location

In fact, Le Havre clay is characterized with a liquid limit $w_l = 79.6\%$, plastic limit $w_p = 44.1\%$ plasticity index $I_p = 35.5\%$, considering an average moisture content of 107.7%, which makes its consistency index negative.

Considering the available data, concluded average grain soil unit weight $\gamma_s = 23.35\,\frac{kN}{m^3}$ and thus for saturated soil its unit weight $\gamma = 13.8k\,\frac{kN}{m^3}$.

Fig. 9. Le Havre site location (France)

3 Measurement Comparison

The used Gothenburg soil data belong to Norconsult Company for soil investigation. Specifically, Vane shear test torque measurements belong to two boreholes named NC 103 and NC 107. Corrected vane shear test measurements obtained at borehole NC115 were used (Fig. 10).

The obtained values of $s_{u_{Tmax}}$ varies from 20 kPa at 6 m depth to 85 kPa at 40 m in average, which makes them exceeding most of the other measurements except for those made with the triaxial test (active mode); In fact these latter are 1.3–2.5 time $s_{u_{Tmax}}$(Figs. 11 and 12).

Comparing $s_{u_{Tmax}}$ and $s_{u_{5.4^\circ}}$, the first are 1.2–12.03 the values of the seconds (Fig. 11).

Particularly $s_{u_{5.4^\circ}}$ varies from 7 kPa at 40 m depth to 18.5 kPa at 3 m depth.

$s_{u_{Tmax}}$ agree well with direct shear test measurements $c_{u_{DSS}}$ as the corrected values from the vane shear test exceed these latter by 1.203 and 1.4 at maximum. This accordance is very noticeable above 12 m depth.

Measurement taken with Triaxial passive test mode: $c_{u_{passive}}$ and fall cone test: $c_{u_{fall\ cone}}$ show lesser accordance with $s_{u_{Tmax}}$, in fact they are 0.8–1.5 and 1.35–2.5 these latter's amount respectively (Fig. 12).

For Tunis soft clay, as illustrated in Fig. 13, the maximal registered torque during vane shear test resulted in a very high s_u measurement above 5 m compared with the rest of the measurements. This was due to soil heterogeneity when reaching the surface. Consequently, it had been chosen to analyze measurements under 6 m depth. As it noticeable $s_{u(Tmax)}$ can be 2–4 time $s_{u_{5.4^\circ}}$.

In fact, s_u measured using maximal torque reaches a maximum value of 18 kPa at 8.6 m, whereas using the proposed method, a maximum value of 5.3 kPa is obtained at only 6.8 m. Attained Minimum values of $s_{u_{Tmax}}$ and $s_{u_{5.4^\circ}}$ were 6.24 and 2.8 kPa respectively at 6.2 m.

Fig. 10. Borehole location (Gothenburg clay)

Correcting the measured undrained shear strength from the maximal torque with Bjerrum coefficient μ equal to 0.94 did not much effect the results, noting that;

$$\mu = 1.1 - 0.005 I_p \in [29.5; 30] \quad (3)$$

The obtained values varies from 5.83 kPa at 6.2 m to 17.1 kPa at 8.6 m.

c_u derived from Skempton correlation (1957) (Eq. 1) resulted in c_u varying from 9.20 kPa at 6.2 m to 12.34 kPa at 8.6 m; 2–3.3 times $s_{u_{5.4^\circ}}$ and 0.6–1.5 $s_{u_{Tmax}}$.

Unlike Gothenburg clay, the range of obtained c_u values still belongs to the used soft clay undrained shear strength ranges.

However, such a comparison is not very credible since from one side the shear vane test available measurements concern a maximum depth of 8.8 m, for a very limited number of boreholes.

In the frame of a running project in Le Havre harbor (passage of a huge diameter pipeline under the harbor's dock), a based vane shear test geotechnical campaign took place recently. Data were used to get s_u profile from the maximum registered torque $s_{u_{Tmax}}$ and torque measured at 5.4° resulting in $s_{u_{5.4^\circ}}$, alike Tunis and Gothenburg clays, and so it is for 20°; $s_{u_{20^\circ}}$, as displayed in Fig. 14. Focus was given on the torque appearing at 20° since it is commonly the failure angle for an over consolidated Kaolin

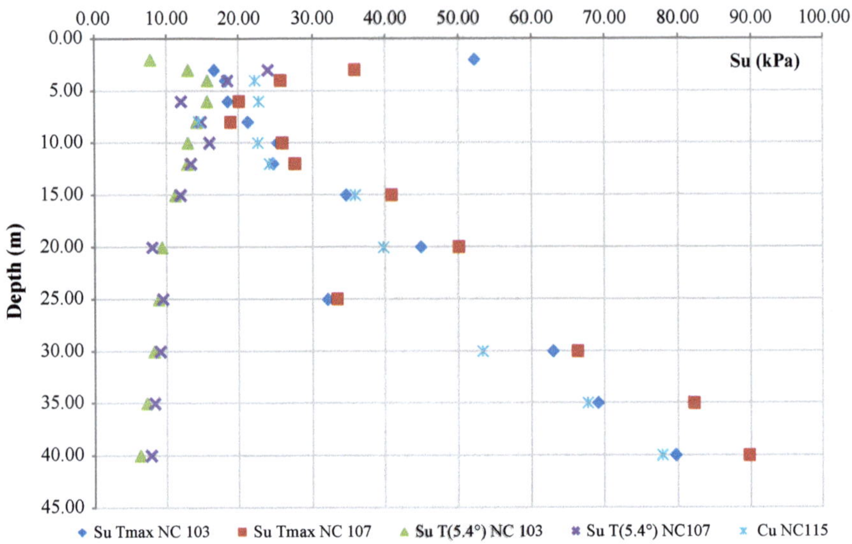

Fig. 11. Short term shear strength profile (Gothenburg clay)

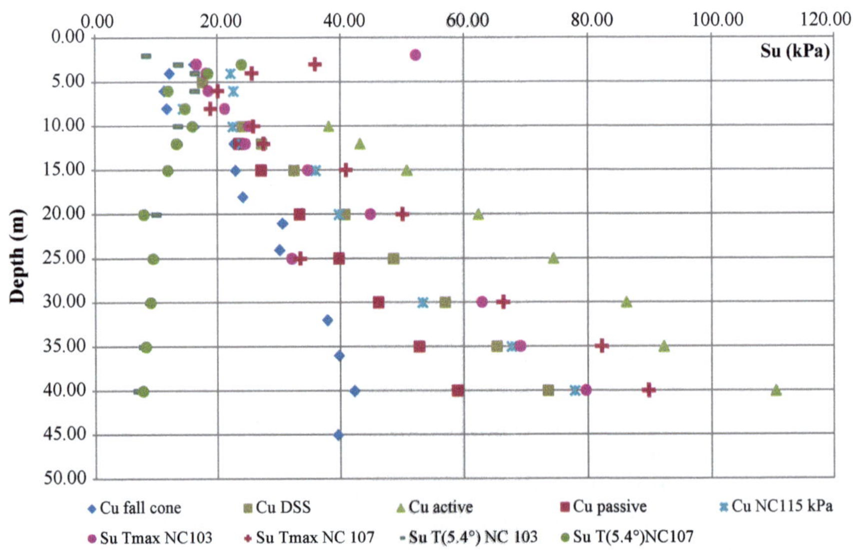

Fig. 12. Comparison between C_u measurements (Gothenburg clay)

(typical soft clay) (Morgenstern and Tchalenko 1967), c_u was also calculated from vane shear test applying SGI (Swedish geotechnical institute) proposed correction μ (Eq. 4) developed in 1984 (Jonsson and Sellin 2012); $c_{u(SGI)}$

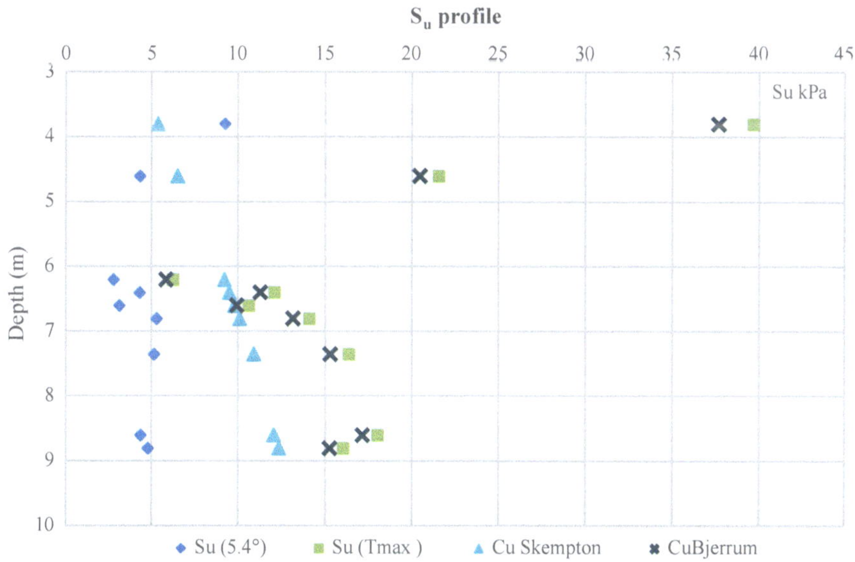

Fig. 13. Short term shear strength profile (Tunis soft clay)

$$\mu = \left(\frac{0.43}{w_l}\right)^{0.45} \geq 0.5 \tag{4}$$

$c_{u(Skepton)}$ was calculated using Skempton correlation.

Alike the previous cases $s_{u_{Tmax}}$ present the most higher obtained values varying from 44 kPa at 13.5 m to 63.13 kPa at 6 m, $s_{u_{Tmax}}$ evaluated at 6.5 m was removed since it reached the maximum torque that can be registered with the vane apparatus at 45.01°, which makes them 1.11 to 18.87 time $s_{u_{5.4°}}$ which varies from 3.31 kPa at 22 m to 11.74 kPa at 5 m.

Applying the SGI proposed correction to vane shear test measurements made these latter 3.37 to.14 time $s_{u_{5.4°}}$; thus no big difference made from $s_{u_{Tmax}}$ case.

$c_{u_{skepton}}$ and $s_{u_{20°}}$ gave the least difference compared with $s_{u_{5.4°}}$ they were from 0.4 to 1.06 and from 2.63 to 4.86 $s_{u_{5.4°}}$ respectively. Particularly $s_{u_{20°}}$ varied from 11.34 kPa at 21.9 m to 32.26 kPa at 6 m (Fig. 14).

Such variation will not demonstrate the true soil behavior regarding the present heterogeneities. This is due the fact that such soil is mainly composed of alluvium deposited within the Seine and it has been since 1517 that this harbor was established to undergo intensive boat traffic.

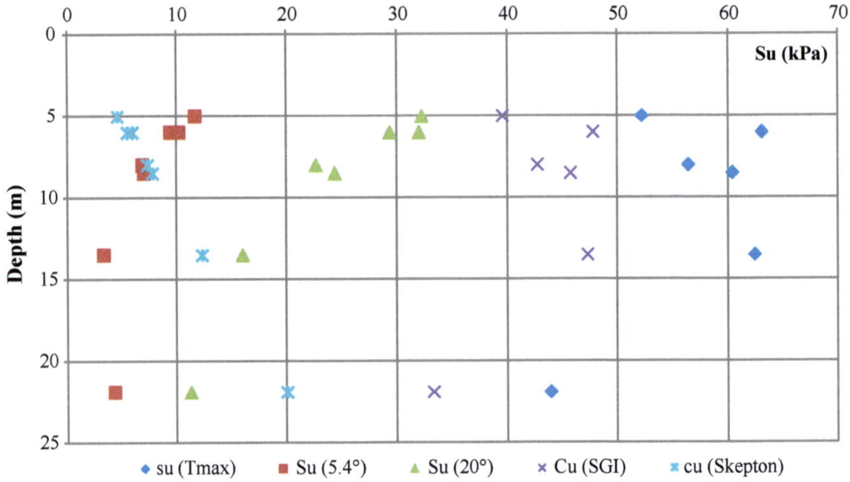

Fig. 14. Short term shear strength profile (Le Havre soft clay)

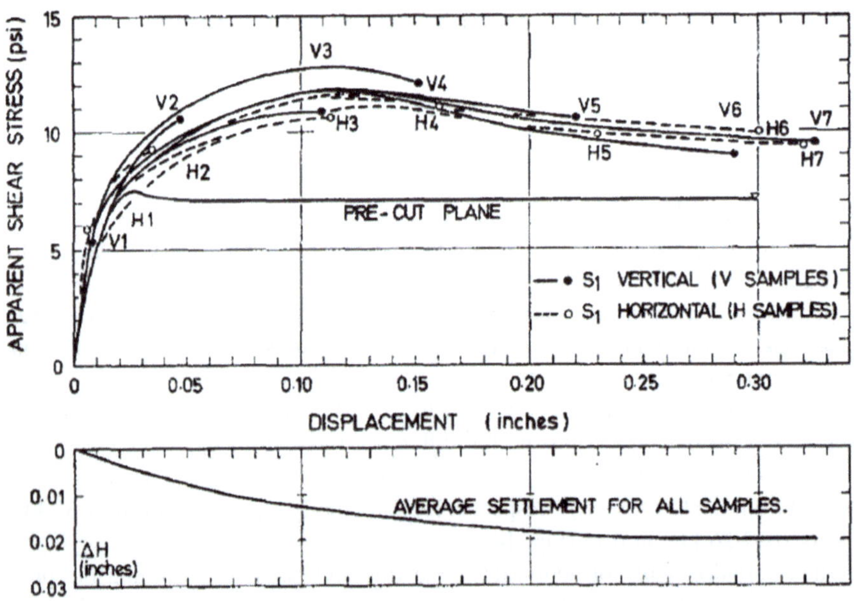

Fig. 15. Shear stress-displacement and settlement curves for drained tests on over-consolidated kaolin. Pre-consolidation load (oedometer) 62.4 lb/sq.in; normal load in shear box test 31.2 lb/sq.in. Morgenster and Tchalenko (1967)

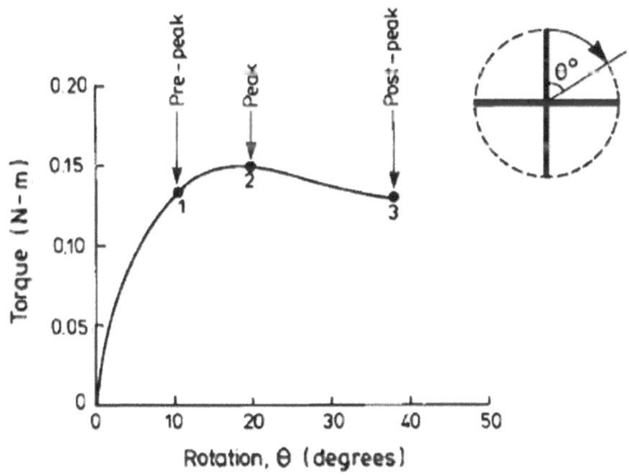

Fig. 16. Development of shear structures in the vane test Figs. 17 and 15 show the structures developed in reconstituted Kaolin at stages 1, 2 and 3. Chandler (1988)

4 Based Rotation Angle Interpretation

To help interpret based rotation angle results we went back to Morgenstern, Tchalenko (1967) and Chandler (1988). In Figs. 18 and 19 resulting deformation during vane shear test and direct shear test were compared along with their evolution through these tests.

Estimated displacements appearing during vane shear test in Fig. 17 for kaolin at 10° matches deformation appearing during the stage v2 (Fig. 19).

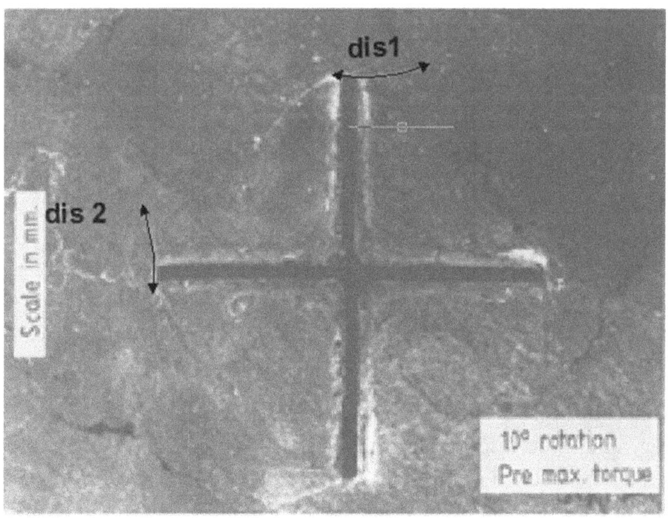

Fig. 17. Pre-peak vane shear structures in kaolin stage 1. Chandler (1988)

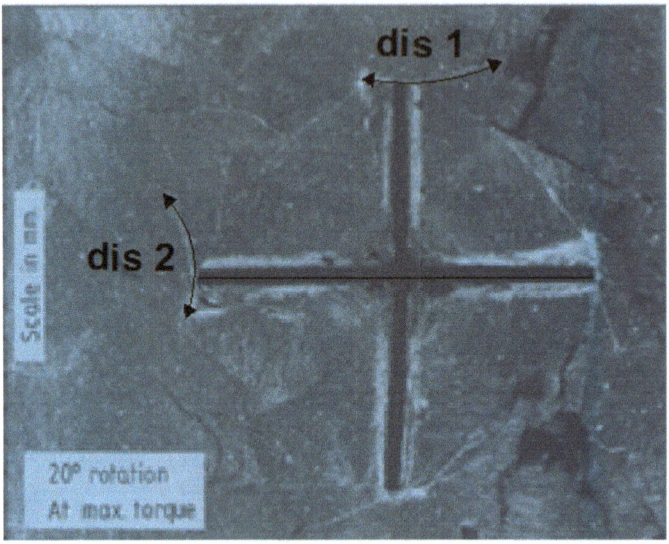

Fig. 18. Vane shear structures at peak shear stress in kaolin stage 2. Chandler (1988)

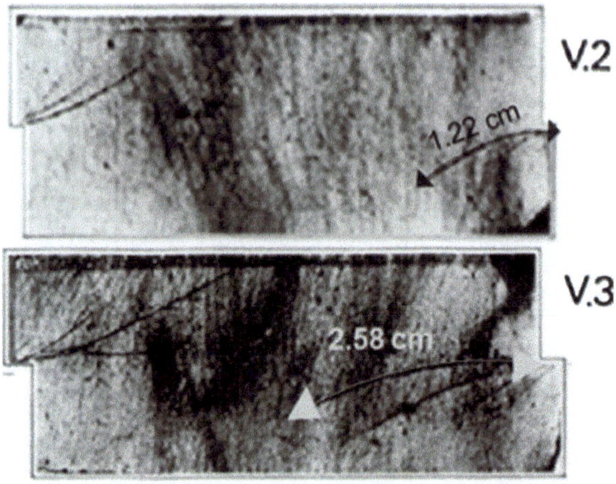

Fig. 19. Sequence of structures in specimens sheared normal to original fabric. Entire longitudinal Sects. 1X2.35 in. Crossed polar (V2 and V3). Morgenstern and Tchalenko (1967)

As illustrated in Figs. 17 and 19 resulting displacements at these states follow the same trajectory and more importantly they present the very same amount; 1.57 cm for vane shear test (see Table 1) and 1.22 cm for direct shear test resulting from a displacement of 1.2 mm as average. These two deformation states belong to the pre-peak curve state (Figs. 15 and 16).

At the peak (Figs. 15 and 16), same thing goes for stage v3, we notice that there is the same amount of displacement 2.25 cm as average for vane shear test and 2.58 cm for direct shear test resulting from a displacement of 2.8 mm (Table 1; Figs. 18 and 19).

Table 1. Resulting displacement during Vane shear test

	Displacement (cm)		
Rotation angle (°)	Dis 1	Dis 2	Average
10	1.66	1.48	1.57
20	2.36	2.15	2.25

These amounts of displacements and deformations still be considerable regarding the volume scale of the studied specimen and the fact that the studied Kaolin which is a typical soft clay was over consolidated.

For Gothenburg soft clay, as illustrated in Fig. 20, the maximum torque was registered at minimum and maximum angles of 8.14° at 4 m and 69.08° at 40 m respectively.

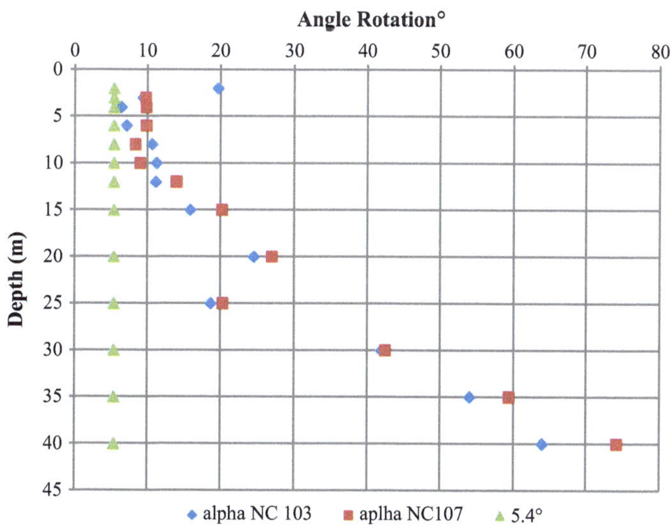

Fig. 20. Rotation angle profile (Gothenburg clay)

It is noticeable that beyond 15 m depth the recorded rotation angle exceeds 20°.

For Tunis soft clay, the obtained maximum torque measurements corresponds to rotation angle varying from 26.6° at 6.8 m to 51.66° at 6.6 m, which means about 5 to 10 time 5.4° (Fig. 21). Considering the rotation amount reached we can assume that the soil had already exceeded little deformation ranges for a little depth which makes it questionable for a soft clay.

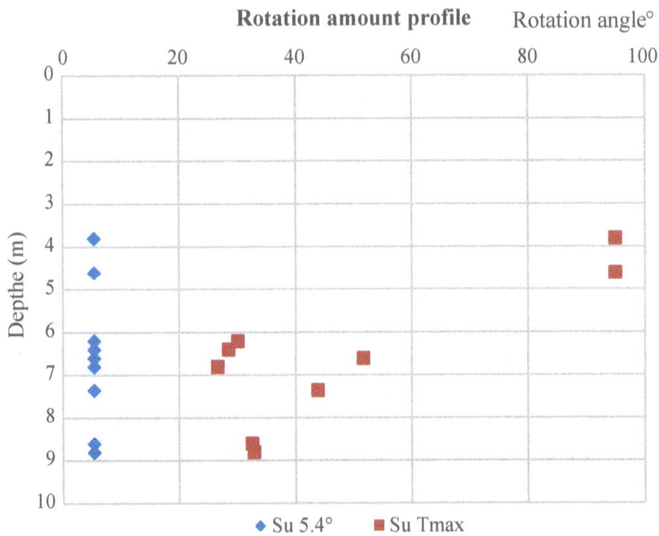

Fig. 21. Rotation angle profile (Tunis soft clay)

For Le Havre soft clay case, rotation angle registered at maximum torque were varying from 37.63° to 69.01° (Fig. 22) which are values exceeding the common rotation angles recorded for soft clay yielding failure when running vane shear test.

5 Based tanλ_{c_u} Interpretation

In this section tan $\lambda_{c_{u_{skempton}}}$ is considered as a reference value particularly for normally consolidated soil, noting that $\tan\lambda_{c_u} = \frac{\Delta_{c_u}}{\Delta_{\sigma_c'}}$.

Regarding this issue the plasticity index is generally constant for a specific soil, and based on Skempton correlation for c_u, tan λ_{c_u} will be a given constant value. For these reasons, it will be possible for normally consolidated soil to evaluate a test efficiency by its ability of giving the most likely uniform profile of tan λ_{c_u}; each point is calculated for a limited soil depth as an average value for a limited soil depth.

As illustrated in Fig. 23 for Gothenburg soft clay, we notice a wide disparity in the tgλ_{c_u} obtained profiles which vary from 0.0016 for tg$\lambda_{s_{u_{5.46°}}}$ $(NC103)$ at 30 m to a maximum value of 1.56 tg$\lambda_{s_{u_{Tmax}}}$ $(NC107)$ at 4 m. Each displayed value is calculated along 1–5 m depth. The obtained profile average is 0.14–0.19.

Calculating tan λ_{c_u} with the proposed method gave a relatively stable variation of tan λ_{s_u} which is more discernible starting from 10 m and 12 m.

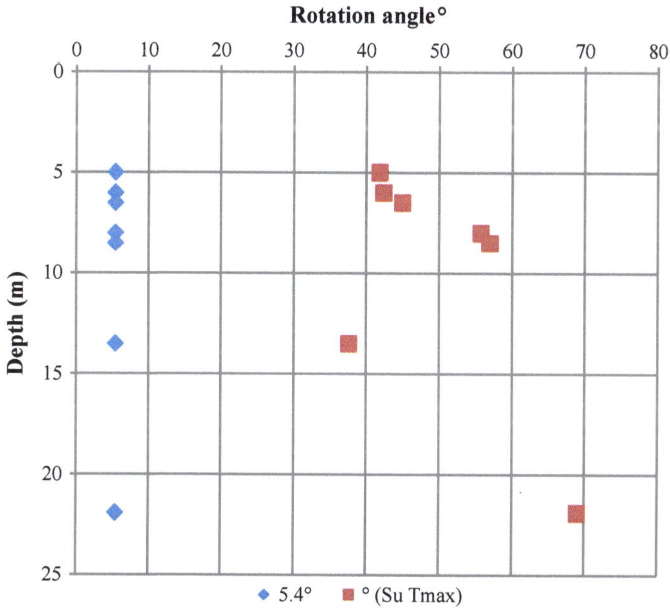

Fig. 22. Rotation angle profile (Le Havre soft clay)

However tan λ_{c_u} calculated with the triaxial test (passive mode) and the direct shear test has shown a more stable variation with 0.19 and 0.25 as average respectively.

Values calculated with the fall cone test, vane shear test without correction and even with those undergoing Bejurrum correction gave a very disperse profile of tan λ_{c_u} (Fig. 23).

The degree of uniformity of tan λ_{s_u} profile can be evaluated by comparing minimum and maximum with each other along with the average value which confirm conclusion from Fig. 23.

For Tunis soft clay $c_{u_{skempton}}$ based calculated tan λ_{c_u} average value is of 0.22. Using $s_{u_{\alpha max}}$ methods for calculating tan λ_{c_u} led to a less dispersed profile. It varies from 0.039 to 1.7 whereas the rest of the methods did not represent any uniform variation to mention (Fig. 24). This interpretation remains relative seeing the lack of data and the shallowness of the investigated depth.

The obtained tan λ_{c_u} profile values of Le Havre soft clay varied in a quite disperse manner.

tan $\lambda_{s_{u(5.4°)}}$ profile was the least scattered and gave the lowest values varying from 0.03 to 0.604 with 0.32 as average value.

tan $\lambda_{s_{u(20°)}}$ profile values were twice as much as these last, varying from 0.06 to 1.4 with 0.69 as average value.

tan $\lambda_{s_{u(Tmax)}}$ and $tan\lambda_{cs_{u(SGI)}}$ profile's resulted in the highest value varying from 0.11 to 2.86 and from 0.083 to 2.16 respectively (Fig. 25).

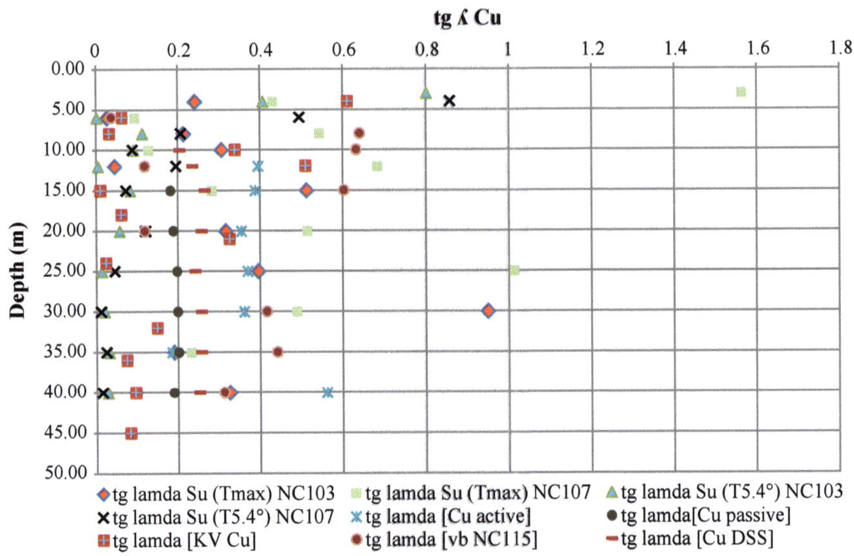

Fig. 23. tan(λ(c_u))profile (Gothenburg clay)

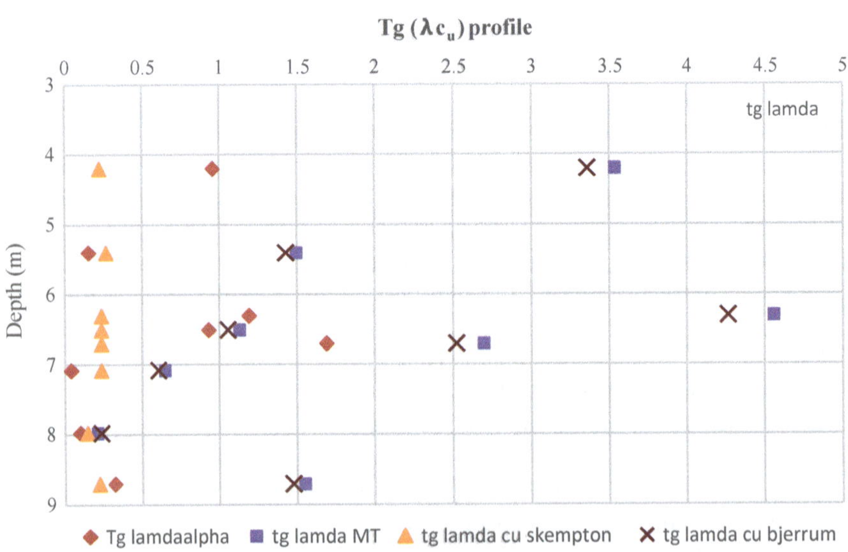

Fig. 24. tan(λ(c_u))profile (Tunis soft clay)

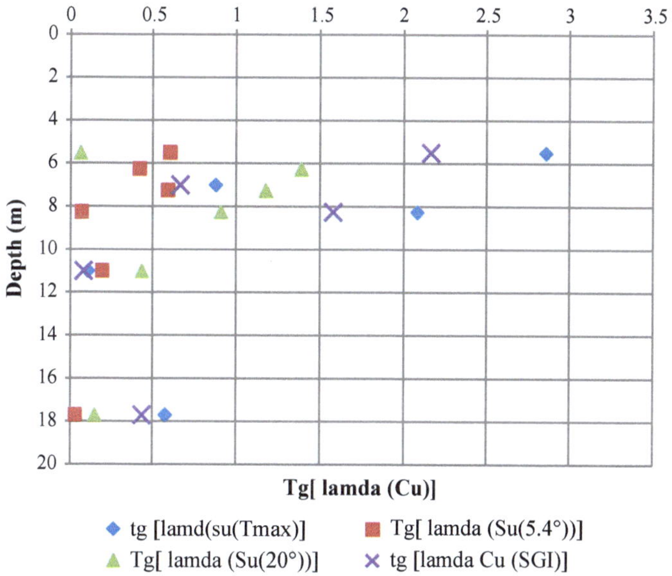

Fig. 25. tan(λ(c_u))profile (Le Havre soft clay)

6 Conclusion

The majority of rotation angles registered for maximal torque during all these test remains unreliable considering consolidation parameter and soil heterogeneities.

Specifically, that heterogeneities inhibits the soil plastic behavior by requiring significant torques and deformations of the tested soil. Thus Bouassida and proposed method will guarantee a soil elastic behavior as mentioned previously.

In fact, by limiting the rotation angle during vane shear test with little deformations range helps avoid short term undrained shear strength overestimation, which represents the most common vane shear test inconvenience.

Chandler, Morgenstern and Tchalenko work gave us an approximation of the angle of rotation of maximum torque for an over consolidated typical soft clay (Kaolin) that does not exceed 20°.

Consequently it will be more reliable to run vane shear test on a normally consolidated Kaolin and observe its behavior to better confirm rotation angle limitation.

If there is no limiting rotation angle vane shear test should be runned on a normally consolidated fine grained soil and a depth limited to 20 m at maximum, otherwise it will lose its signification.

Acknowledgements. The authors address sincere thanks to Goran Sallfors for his help and kind cooperation. His help was very crucial and gave much value to this study.

The help provided by the "Hydrogeotechnique" and Cerema Company (France) is also appreciated by the authors.

References

Bouassida, M., Boussetta, S.: On the determination of vane shear strength of soft soils. In: Proceedings of the 12th African Regional Conference on Soil Mechanics and Geotechnical Engineering, Durban (South Africa), pp. 285–291, 24th–27th Oct (1999)

Chandler, R. J.: The in-situ measurement of the undrained shear strength of clays using the field vane. In: Richards, A.F. (ed.) Vane Shear Strength Testing in Soils, Field and Laboratory Studies, ASTM STP 1014, pp. 13–44. American Society for Testing and Materials, Philadelphia (1988)

Costet, J., Sanglerat, G.: Cours de mécanique des sols, tome 2, edition Dunod (1983)

Mahmoud, M.: Vane testing in soft clays. Report of British Geotechnical Society's, informal discussion held at the Institution of Civil Engineers, 8th June 1988 (1988)

Morgenstern, N.R., Tchalenko, J.S.: Microscopic structures in kaolin subjected to direct shear. Géotechnique **17**, 309–328 (1967)

Roy, M., Leblanc, A.: Facteurs influençant la mesure au scissomètre, Bull. liaison P.et Ch, 144 Juil.-Août 1986 –réf, 311 2 (1986)

Toalen, J-P., Duron, C.,Hagnéré, C.: Traitement de la vase de l'estuaire de la seine. In: SESSION V: Dragages, Valorisation des matériaux marins,Matériaux.Article n° 48 Actes IVèmes JNGCGC, Dinard 1996 (1996)

Touiti, L., Boussetta, S., Bouassida, M.: Geotechnical study of tunis Soft Clay. In: Conference of Soil 14th ARC SMGE, 26th–28th Nov, Yaounde (2007)

Issues in Interpretation of Constant Rate of Strain Consolidation Test Data

Gyan Vikash(✉) and Ramyasri Rachamadugu

Department of Civil Engineering, School of Engineering, Shiv Nadar University, Gautam Buddha Nagar, Noida, Uttar Pradesh, India
{gyan.vikash, rr783}@snu.edu.in

Abstract. Constant rate of strain (CRS) consolidation test is a fast test method to characterize consolidation behavior of fine grained soils. In this test method, the test specimen is deformed at a constant rate of deformation, and pore water pressure at the base of the test specimen and axial reaction load are measured at successive interval during the test. This test method has several advantages over traditional incremental loading consolidation test. In the present research, a series of CRS consolidation tests have been performed on reconstituted samples of eight different clayey soils with different plasticity indices. The suitable strain rate at which the test is to be performed has been decided considering the criterion given in the literature. The test data obtained from these tests have been analyzed through the method of interpretation given in ASTM: D4186-06 (Standard Test Method for One-Dimensional Consolidation Properties ofSaturated Cohesive Soils Using Controlled-Strain Loading. 2008). The present study shows that the existing method of interpretation yields unreliable results for a significant duration at early stage of the test despite of performing the test at recommended strain rate. Analysis of present experimental data and some of the available recommendations to select suitable strain rate indicates that the theory for CRS consolidation which forms the basis for the existing method of interpretation is unable to describe rapid evolution of pore-water pressure at early stage of the test. Because of this, the existing method of interpretation fails to interpret the consolidation parameters accurately throughout the test.

1 Introduction

Characterization of consolidation behavior of fine grained soils is primarily required to predict amount of settlement and rate of settlement of soil layer beneath the structures. Conventionally, it is done through one-dimensional Incremental Loading (IL) consolidation test. In IL consolidation test, load increment ratio is usually kept one, which results in too large spacing between the data points to determine some of the consolidation parameters accurately. In addition to this, IL consolidation test is very time consuming as it takes somewhere around 7–12 days to perform one test. A complete project may have around 50–100 such samples to characterize, which makes it a formidable task sometimes to perform IL consolidation test considering the time line of the project and the amount of effort required. Hamilton and Crawford (1959) introduced a fast test method to characterize the consolidation behavior of soil which is

D. Chen et al. (eds.), *Advancements on Sustainable Civil Infrastructures*,
Sustainable Civil Infrastructures, https://doi.org/10.1007/978-3-319-96241-2_6

known as Constant Rate of Strain (CRS) consolidation test. In this test method, the soil specimen is deformed at a specified axial deformation rate, and reaction load, deformation and excess pore pressure at the base of the specimen are measured during the test at successive intervals. CRS consolidation test takes relatively very less time (few hours) to characterize the consolidation behavior of fine grained soils. The testing method is simple enough for any laboratory to adopt. This test method provides many more data points in comparison to IL consolidation test. Several theories have been proposed to interpret the CRS consolidation test data (Smith and Wahls 1969; Wissa et al. 1971; Yoshikuni et al. 1995). Among these theories, Wissa et al.'s theory is more comprehensive and it has also been incorporated in ASTM standard D4186-06: Standard Test Method for One-Dimensional Consolidation Properties of Saturated Cohesive Soils Using Controlled-Strain Loading.

Although the testing method and the method of interpretation for CRS consolidation test data are well established, there are several issues which need to be addressed to make CRS consolidation test as a standard laboratory test to characterize consolidation behavior of fine grained soils. These issues are: (1) unrealistic interpretation of the results for a significant duration at early stage of the test, (2) unavailability of the method of interpretation to interpret the test data of unloading-reloading stages, (3) selection of suitable strain rate, and (4) interpretation of the test data of soft soils, where small strain theory does not applicable. Because of these issues, CRS consolidation test is not being widely used in practice despite its advantages.

The present study evaluates issue of unrealistic interpretation of the consolidation parameters at early stage of the test. In order to facilitates this study, reconstituted samples of eight different soils having different plasticity indices were prepared by slurry consolidation technique. The test specimens obtained from the bulk samples prepared by slurry consolidation technique were consolidated at a CRS. The suitable strain rate was selected considering the recommendation given in the literature. The test data was interpreted using the method of interpretation given in ASTM: D4186-06 (2008). The interpreted results of all the soils considered in the present study show that the existing method of interpretation yields unrealistic results for significant duration at early stage of the test despite of performing the tests at recommended strain rate. Further, some of the recommendations on selection of suitable strain rate to perform the CRS consolidation test suggested by many researchers in the past have been analyzed. It is generally believed that performing the test at inappropriate strain rate is the reason behind issue of unrealistic interpretation at early stage of the test. The present analysis however indicates that the issue of unrealistic interpretation at early stage of the test does not occur because of inappropriate strain rate; it rather occurs because of inability of the theory forming basis of the existing method of interpretation in defining the early stage mechanism of evolution of pore water pressure.

2 Experimental Study

2.1 Material Properties

The present experimental study was carried out on reconstituted samples of eight different soils having different plasticity indices. Table 1 shows index properties of the soils considered in the present study. For convenience, the samples of different soils have been identified as the sample ID F1–F8.

2.2 Experimental Setup

A CRS consolidation testing apparatus was designed and fabricated for the present research. Figure 1 shows details of the CRS consolidation test setup. CRS consolidation cell and loading frame are two main components of CRS consolidation testing apparatus. CRS consolidation cell consists of base and transparent perspex cell with aluminium top. Base, made of aluminium, has mainly two valves; one is used to connect a pressure transducer to measure pore water pressure at the bottom of the specimen and another was used to fill water in the cell. A stainless steel consolidation ring of internal diameter 75 mm and height 25 mm was used to hold the test specimen. An arrangement was made to attach the consolidation ring to the base of the consolidation cell as shown in Fig. 1b. A pressure line was connected at the top of the consolidation cell to apply the desired amount of back pressure in the test specimen through surrounding water. A standard triaxial loading frame with compression machine of maximum rated capacity of 20 kN and loading speed of range 0.0001–9 mm/min was used to load the specimen at a constant rate of deformation. During the test, pore pressure at the bottom was measured by a transducer connected to the base of the consolidation cell, load was measured using the load cell of capacity 10 kN, and the imposed displacement was measured by a linear voltage displacement transducer (LVDT). All the transducers were connected to an automatic data acquisition unit to allow recording during the test.

Table 1. Index properties of the soils considered in the present research

Sample ID	Initial void ratio	Water content (%)	Liquid limit (%)	Plastic limit (%)	Plasticity index (%)
F1	1.17	45	65	30	35
F2	1.08	41	57	29	28
F3	1.02	39	48	28	20
F4	0.94	36	35	26	9
F5	0.84	32	32	24	8
F6	0.78	30	26	22	4
F7	0.86	33	41	25	16
F8	0.90	35	45	19	26

2.3 Preparation of Bulk Samples

Slurry consolidometer of acrylic material were fabricated to prepare the bulk samples. Figure 2 shows its overall assembly and details of all the components. Initially, a slurry

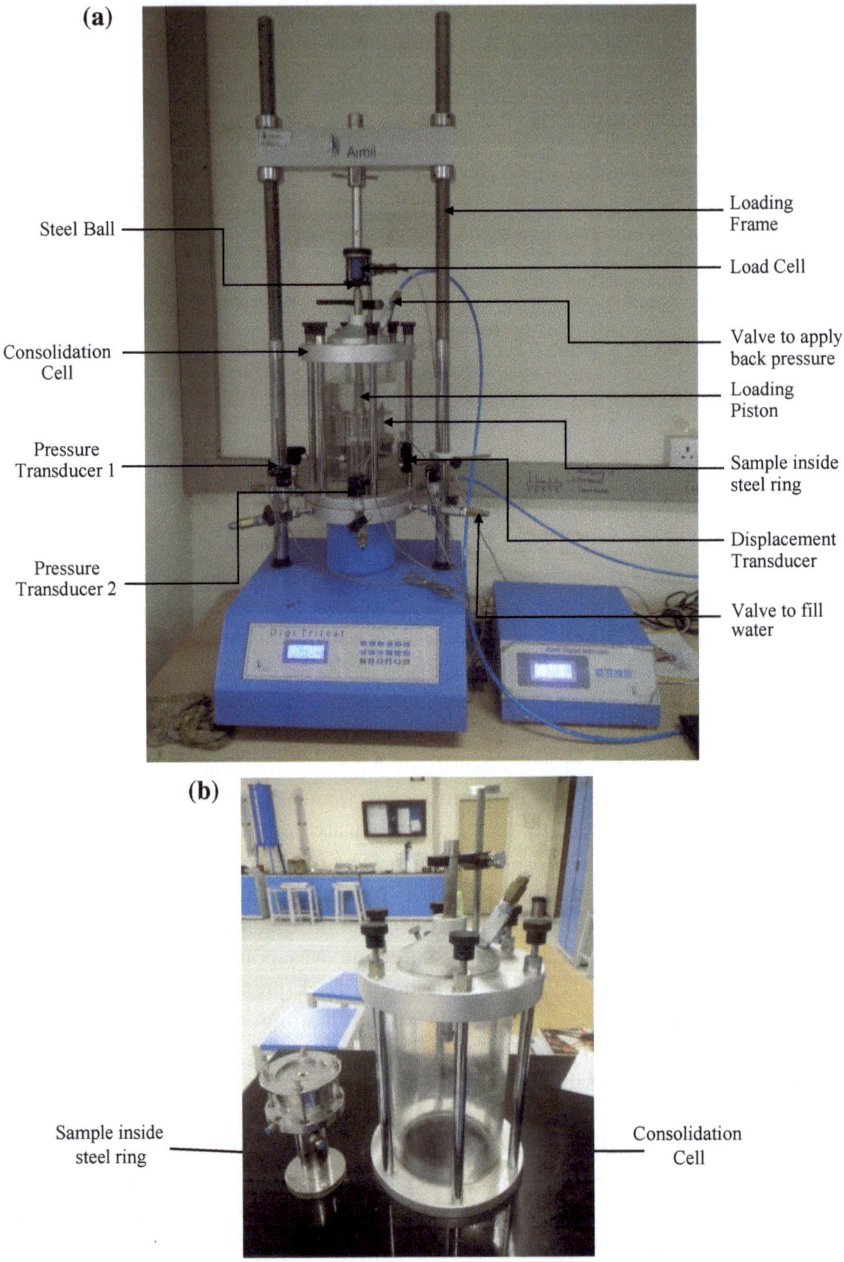

Fig. 1. **a** CRS consolidation testing apparatus. **b** Consolidation cell and consolidation ring. **c** Details of consolidation ring

(c)

Fig. 1. (*continued*)

of the soil was prepared by mixing 2.5 times water by weight to oven dried soil powder. Much care was taken to insure the uniformity of the prepared samples. This slurry was then carefully poured into an acrylic cylinder, as shown in Fig. 2. The sample was then kept for consolidation pressure of 200 kPa by applying vertical pressure gradually from the pneumatic piston at the top. Drainage of pore water from the sample takes place through top and bottom of the sample. The sample was kept for consolidation until no further displacement of piston was observed. Thereafter, the test specimens were taken out from the bulk sample once the consolidation was done.

A wire saw was used to cut the bulk sample in order to obtain the test specimen for consolidation test. The consolidation steel ring with a grease gel coating on its inner circumference was inserted into the part of the bulk sample very carefully. Wire saw was used to remove excess soil and get a smooth surface on both sides of the ring. Consolidation ring was then assembled, as shown in Fig. 1b, and kept inside the consolidation cell. Distilled and deionized water was filled inside the consolidation cell through a valve present at the base of the frame to an appropriate level so that the consolidation ring is totally submerged. Appropriate back pressure is applied on the test sample through air pressure line. The setup is then left for 30 min to allow proper saturation of the sample and to equilibrate the applied back pressure. Meanwhile, the load cell and displacement transducer were attached with the setup to record the data. Once the saturation was attained, the test specimen was consolidated at deformation rate of 0.02 mm/min.

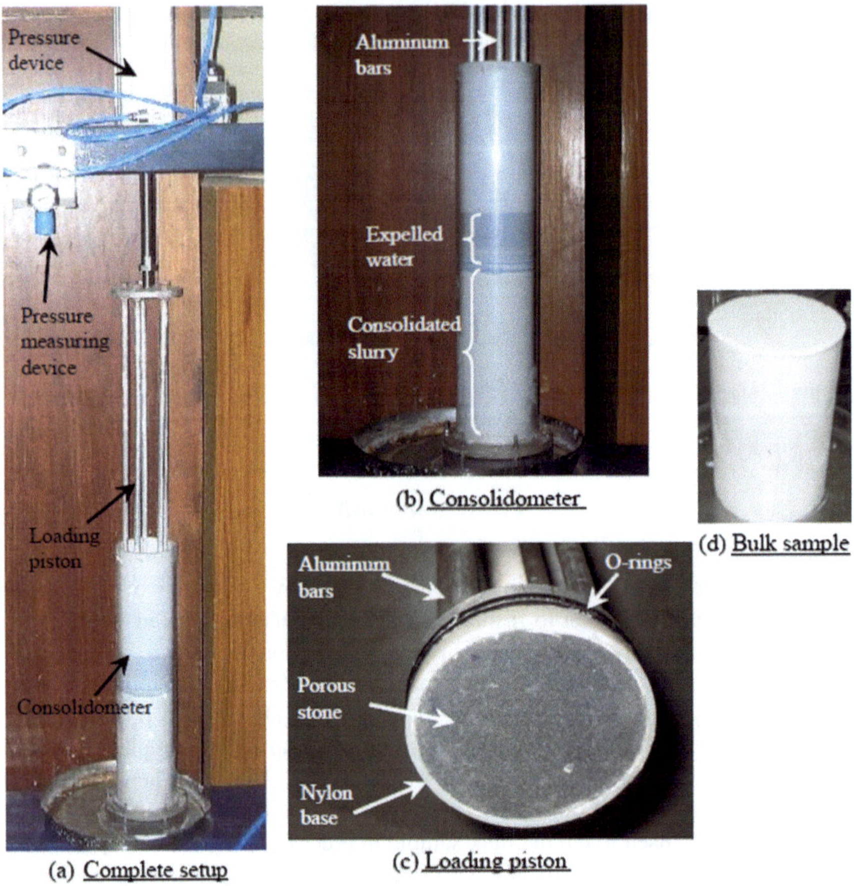

Fig. 2. Slurry consolidation setup. **a** Complete setup during slurry consolidation, **b** water expelled in the consolidometer during consolidation, **c** details of loading piston, **d** extruded bulk sample after completion of slurry consolidation

3 Method of Interpretation of Test Data

Wissa et al. (1971) proposed a comprehensive theory for CRS consolidation which forms the basis of method of interpretation given in ASTM standard D4186-06: Standard Test Method for One-Dimensional Consolidation Properties of Saturated Cohesive Soils Using Controlled-Strain Loading. They derived a solution for CRS consolidation which consists of transient state and steady state components. According to Wissa et al., transient condition is developed in the test specimen immediately as the piston is set in motion and dissipates after sufficient duration. Thereafter, the steady state condition prevails within the test specimen. They formulated the following governing differential equation for CRS consolidation

$$c_v \frac{\partial^2 \epsilon}{\partial z^2} = \frac{\partial \epsilon}{\partial t} \tag{1}$$

where, c_v is coefficient of consolidation, ϵ is strain at any location z within the specimen at time t during the test.The solution of Eq. (1) is a function of non-dimensional depth (Z) and time factor (T_v), given as

$$\epsilon(Z, T_v) = rt\left[1 + \frac{1}{6T_v}(2 - 6Z + 3Z^2) - T(Z, T_v)\right] \tag{2a}$$

$$T(Z, T_v) = \frac{2}{\pi^2 T_v}\sum_{n=1}^{\infty}\frac{\cos n\pi Z}{n^2} exp\left(-n^2\pi^2 T_v\right) \tag{2b}$$

where $Z = z/H$ in which H is the specimen thickness; $T_v = c_v t/H^2$, r is the rate of strain, and n is an integer. The solution for Eq. (1) consists of two parts, (2a) and (2b). The first part is the steady state component and the second part is the transient state component. They derived the expressions for σ', k, and c_v for steady state condition. The steady state component given in Eq. (2a), can be written in dimensional form as,

$$\epsilon(z, t) = rt + \frac{rH^2}{C_v}\left[\frac{1}{6}\left[\frac{1}{3}\left(\frac{z^2}{H} - 6\frac{z}{H} + 2\right)\right]\right] \tag{3}$$

In order to derive an expression for σ', k, and c_v, they assumed coefficient of volume compressibility (m_v) to be constant in subsequent time increment. For linear stress–strain relationship, they obtained the expression for m_v as,

$$m_v = \frac{rH^2}{2C_v u_b} \tag{4}$$

where, u_b is pore water pressure at the base measured during the test.

$$C_v = \frac{k}{m_v \gamma_w} \tag{5}$$

Hence, coefficient of permeability can be written as

$$k = \frac{rH^2\gamma_w}{2u_b} \tag{6}$$

The expression for vertical effective stress, σ' can be written as

$$\sigma' = \sigma - \frac{2}{3}u_b \tag{7}$$

In order to determine the time when the transient state which develops at the beginning of the test dissipates, they introduced a function $f_3(T_v)$, which is defined as

$$f_3(T_v) = \frac{\epsilon(1, T_v)}{\epsilon(0, T_v)} \tag{8}$$

Further, f_3 can be expressed as

$$f_3 = \frac{(\sigma - u_b) - \sigma_{t=0}}{\sigma - \sigma_{t=0}} \tag{9}$$

They suggested that the transient state completely dissipates for $f_3 \geq 0.4$ for linear material.

4 Interpretation of the Test Data

Evaluation of complete dissipation of transience is important to determine to interpret the CRS consolidation test data as the method of interpretation is available only for steady state condition. To determine the test data corresponding to steady state condition, f_3 as shown in Eq. (9) was determined for all the test specimens, F1 to F8. It was observed that f_3 was greater than 0.4 for all the test data and it ranged from 0.7 to 0.8. This indicates that, according to ASTM: D4186-06 (2008), transient state does not exist in any of the test specimens. It can be therefore considered that the steady state condition prevails from beginning of the test in all the test specimens, and hence the test data of all the specimens F1 to F8 measured from beginning of the test can be used for interpretation of consolidation parameters.

According to recommendations given in the literature CRS consolidation test should be performed at the strain rate which yields the value of $R_u = u_b/\sigma$ throughout the test less than 0.5.

Figure 3 shows the variation of pore water pressure ratio, R_u with total stress for the specimens F1 to F8. It can be observed from this figure that the value of R_u is less than 0.5 for all the test specimens. This indicates that the deformation rate of 0.02 mm/min considered in the present experimental study satisfying the recommendation given in the literature for selection of suitable strain rate. From this figure, it can also be observed that R_u is higher for the specimen with lower void ratio (Sample F7 has highest value of R_u, and Sample F1 has lowest). In addition to this, Fig. 3 shows that R_u of all the test specimens decreases asymptotically as the test progresses.

Figure 4 shows the variation of interpreted coefficient of permeability (k) with void ratio (e) of the specimens F1 to F8 in semi-logarithmic space. The value of k was determined using the formula as shown in Eq. (6). From this figure, it can be readily observed that for all the test specimens, k decreases rapidly with e at early stage, which has been shown as OA in Fig. 4, whereas variation of k with e is moderate at a later stage (AB). In the stage OA, e decreases by a very small amount for, e.g., 1.17–1.12 for the specimen F1, and 0.78–0.75 for the specimen F6, whereas k decreases by about 300 folds. This much amount of variation in k for a very small change in e can be hardly

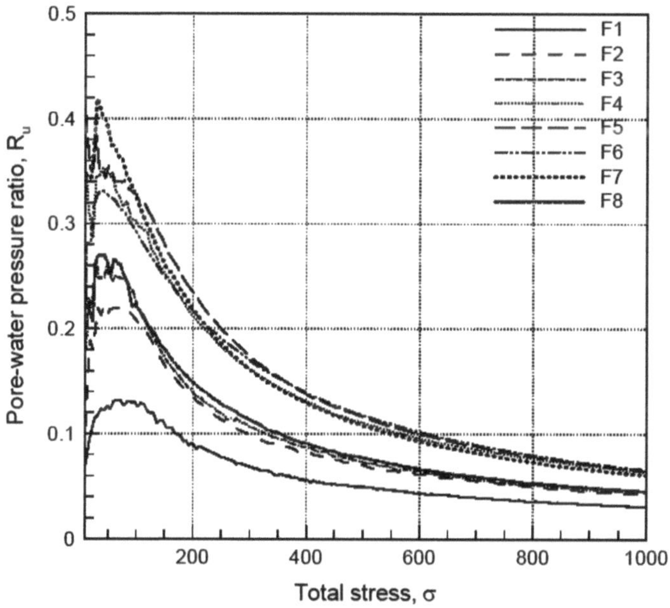

Fig. 3. Variation of pore-water pressure ratio, R_u, with the total stress, σ for all the test specimens F1 to F8

observed in reality. The observed variation of k with e indicates that the existing method of interpretation provides unrealistic interpretation at early stage of the test, however it interprets reasonably in the later stage. In addition to this, Fig. 4 shows that e-log k curve of all the specimens follows bilinear curve, it is however well established fact that e-log k curve follows linear relationship irrespective of testing conditions. Therefore, bilinear trend of e-log k curve imposes a question that which linear part of the curve should be considered as true representation of variation of k with e. The observed trend suggests that the test data of the later stage can be considered to interpret the variation of k with e. However, at the same time, it points out a limitation of the available method of interpretation to interpret the early stage test data.

Yoshikuni et al. (1995) suggested that, in CRS consolidation test, the consolidation parameters should be interpreted using the test data after the time from where the slope of the curve between pore-water pressure at the base, u_b, versus t become gentler. According to Yoshikuni et al. (1995), transience is significant in the time domain where the curve between u_b versus t is steeper. Figure 5 shows variation of u_b, with t. In this figure, the time up to which the curve between u_b versus t is steeper, is denoted by t_{tr}. From this figure, it can be observed that the value of t_{tr} for the specimen F1, F2, and F3 is about 30 min, whereas it is about 40 min for the specimens F4, F5, F6, F7, and F8. From this figure, it can be also observed that the total time required to complete the test lies between 80 min for the specimen F6 to 180 min for the specimen F1. Total time required to complete the test is corresponding to the stress level of 1000 kPa, as in the present experimental study all the test specimens are monotonically loaded up to

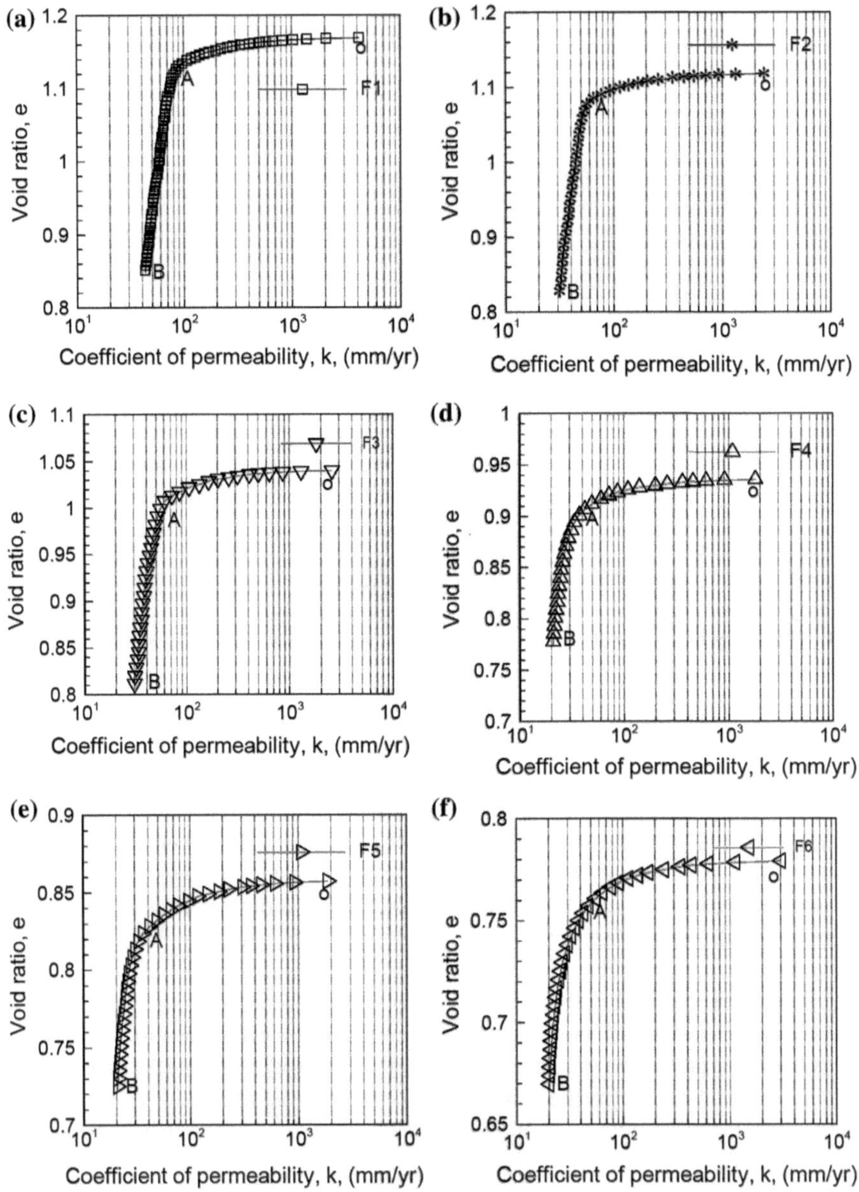

Fig. 4. Variation of *k* with void ratio for all the test specimens F1 to F8

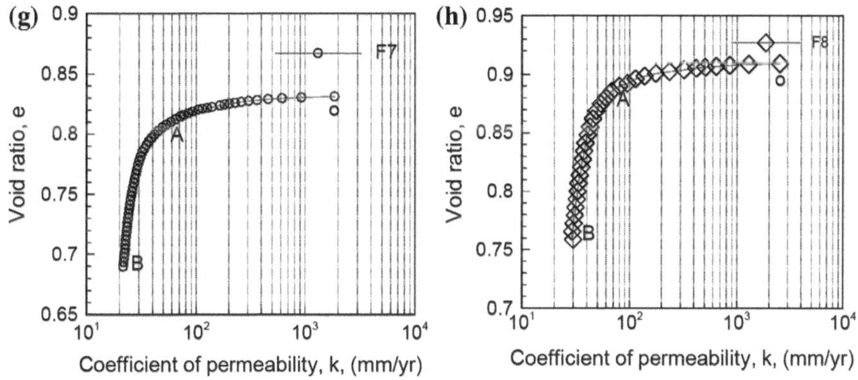

Fig. 4. (*continued*)

1000 kPa. From Fig. 5, it can be deduce that the time by which the curve between u_b versus t is steeper, t_{tr}, is significant with respect to the total time required to complete the test. It is interesting to note that the region OA, as shown in Fig. 4, where unrealistic interpretation occurs, lies in the time domain less than t_{tr}. This indicates that the interpretation of the test results through the existing method of interpretation is unreliable in the stage where the slope of the curve between u_b versus t is steeper, whereas it is somewhat reasonable in the stage where the slope of the curve between u_b versus t is gentler.

5 Analysis of Some Recommendations on Selection of Suitable Strain Rate

Issue of unrealistic interpretation for significant duration at early stage of CRS consolidation test was also reported by many researchers in the past (Smith and Wahls 1969; Wissa et al. 1971; Gorman et al. 1978; Armour and Drnevich 1986; Lee et al. 1993; Sheahan and Watters 1997). They reported that at early stage of the test, the existing method of interpretation provides unrealistic interpretation for k and c_v, and shows a significant disparity between e versus σ' relationship obtained from CRS consolidation test and that obtained from IL consolidation test. They argued that it might occur because of selection of inappropriate strain rate. They thereby suggested various recommendations on selection of suitable strain rate at which test can be performed. Smith and Wahls (1969) suggested that the strain rate should be such that the pore-water pressure ratio, R_u, throughout the test should be less than 0.5. Wissa et al. (1971) suggested that the selection of proper strain rate depends on the parameters required. If compression response is required, the test should be performed at a strain rate such that u_b remains close to zero. If k and c_v are required, a strain rate is set such that the finite hydraulic gradient develops within the specimen. Further, Lee et al. (1993) suggested that the rate of strain is acceptable if the coefficient of consolidation derived from an analysis of the drained face equals that from the undrained face. They

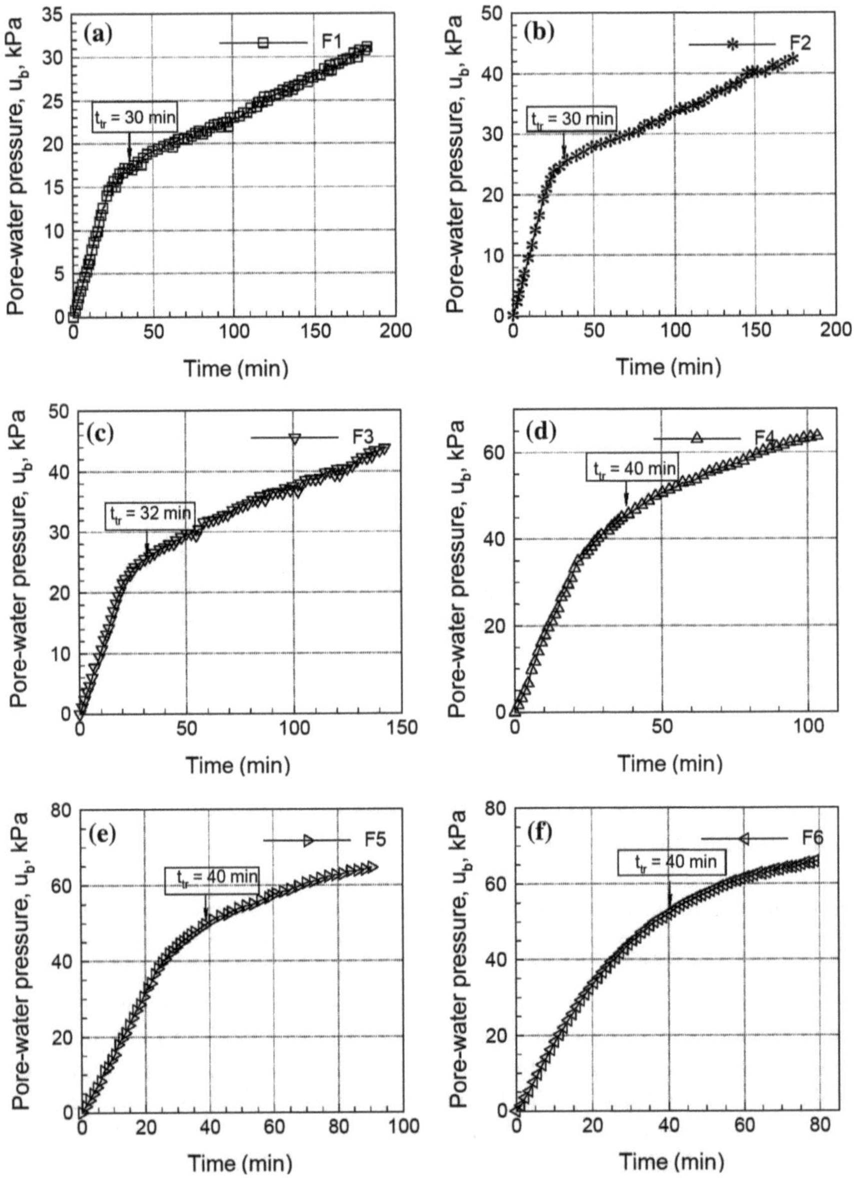

Fig. 5. Variation of pore-water pressure at the base of the test specimen during the test for all the test specimens F1 to F8

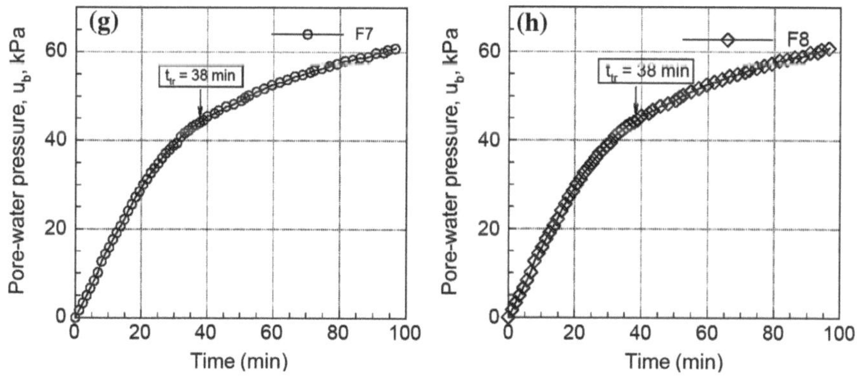

Fig. 5. (*continued*)

found that CRS consolidation test results and IL consolidation test results are nearly the same at this strain rate. From Wissa et al.'s theory, they derived the following expressions for c_v at the drained and undrained face. c_v at the drained face is given by

$$c_{v\,drained} = \frac{h^2}{2u_b}\frac{\Delta\sigma'_d}{\Delta t} \tag{10}$$

where, h is the current thickness of the specimen, $\Delta\sigma'_d$ is the change in the effective stress at the drained face in time Δt. c_v at the undrained face is given by

$$c_{v\,undrained} = \frac{h^2}{2u_b}\frac{\Delta\sigma'_{ud}}{\Delta t} \tag{11}$$

where $\Delta\sigma'_{ud}$ is the change in effective stress at the undrained face in time Δt. According to Lee et al. (1993), the test should be performed at a strain rate at which the two coefficients, given by Eqs. (10) and (11), converges.

The criteria given by Lee et al. (1993) is further analyzed herein to gain insight on the concept behind it. Since the excess pore-water pressure is zero at the drained face, therefore $\Delta\sigma'_d$ will be equal to $\Delta\sigma$. $\Delta\sigma'_{ud}$ will be equal to $\Delta\sigma - \Delta u_b$ since the excess pore-water pressure at the undrained face is u_b. Substituting $\Delta\sigma'_d = \Delta\sigma$ and $\Delta\sigma'_{ud} = \Delta\sigma - \Delta u_b$ into Eqs. (10) and (11), and then subtracting Eq. (10) from Eq. (11), following expression is obtained,

$$c_{v\,drained} - c_{v\,undrained} = \frac{h^2}{2u_b}\frac{\Delta u_b}{\Delta t} \tag{12}$$

Equation (12) depicts that the coefficient of consolidation at the drained face, $c_{v\,drained}$, and at the undrained face, $c_{v\,undrained}$, will converge only when $\Delta u_b/\Delta t$ will become close to zero.

Lesser disparity between these two coefficients, which is generally observed at lower strain rates, is due to the fact that insignificant amount of u_b evolves in the test performed at lower strain rate. As Fig. 5 shows that at early stage of the test, slope of the curve between u_b and t is steeper, hence $\Delta u_b/\Delta t$ cannot be close to be zero at this stage. However, $\Delta u_b/\Delta t$ can be considered close to be zero at later stage as the slope of the curve between u_b and t is gentler at this stage. From the present experimental study, it can be inferred that the existing method of interpretation does not yield reliable results in the stage where u_b versus t curve is steeper, whereas it gives reasonable results in the stage where u_b versus t curve is gentler. This indicates that the existing method of interpretation fails to interpret accurately the test data corresponding to rapid varying u_b with time during the test. It is important to note that, according to ASTM standard, all the test data including the test data corresponding to rapid varying u_b with time correspond to the steady state condition. And hence failure of the method of interpretation to interpret the test data reasonably indicates inability of the method of interpretation to define mechanism of rapid evolution of u_b at early stage of the test. It appears that the transient state is significant in the stage where u_b varies rapidly with time and it becomes insignificant in the stage where variation of u_b with time is gentler. In the light of observed experimental results and present analysis of criterion for selection of suitable strain rate, criterion given in ASTM standard (ASTM: D4186-06, 2008) to define attenuation of transience seems to be incorrect. Therefore, theory behind the method of interpretation of CRS consolidation test data need to be relooked in order to understand the mechanism of evolution of pore-water pressure distribution during the test. The present study indicates that it is not the strain rate which causes issue of unrealistic interpretation, it is nevertheless inability of the theory formulating the method of interpretation in defining rapid variation of u_b with time t at early stage of the test. Mechanism explaining rapid evolution of u_b with time t at early stage need to be incorporated in the theory for CRS consolidation to resolve the issue of unrealistic interpretation at early stage of the test.

6 Conclusions

In the present study, issue of unrealistic interpretation at early stage of CRS consolidation test was evaluated. A series of CRS consolidation tests were performed on reconstituted samples of eight different clayey soils with different plasticity indices. Bulk samples were prepared by slurry consolidation technique in which slurry of the soils were consolidated at 200 kPa. The test specimen for CRS consolidation were taken out from the bulk samples and then it was consolidated to a stress level of 1000 kPa at a constant rate of deformation of 0.02 mm/min. The test data was interpreted using the method of interpretation given in ASTM standard D4186-06. The results obtained from this experimental study indicates that the available method of interpretation yields unrealistic results for significant duration at early stage of the test despite of performing the test at suitable strain rate. Further, some of the recommendations on selection of suitable strain rate available in literature were analyzed to gain insight on factors deciding selection of appropriate strain rate for CRS consolidation test. The present analysis and the observed experimental results indicate that the issue of unrealistic interpretation occurs due to inability of the theory which forms the basis for available method of interpretation to describe the early stage mechanism of evolution of pore-water pressure. Therefore, in order to resolve the issue of unrealistic interpretation, early state mechanism during CRS consolidation needs to understand in more detail as well as it should be incorporated in CRS consolidation theory.

Acknowledgements. Financial supports from Department of Science and Technology (DST) India through grants SERB/FTP/ETA-0041/2014 is gratefully acknowledged. Any opinions, findings, and conclusions or recommendations expressed in this material are those of authors and do not necessarily reflects the views of DST.

References

Armour, D.W., Drnevich, V.P.: Improved techniques for the constant rate of strain consolidation test. In: Consolidation of Soil, STP 892, pp. 170–183. American Society for Testing and Materials, Philadelphia (1986)

ASTM: D4186-06: Standard Test Method for One-Dimensional Consolidation Properties of Saturated Cohesive Soils Using Controlled-Strain Loading (2008)

Crawford, C.B.: On the importance of rate of strain in the consolidation test. Geotech. Test. J. **11** (1), 60–62 (1988)

Fox, P.J., Pu, H., Chritian, J.T.: Evaluation of data analysis methods for the CRS consolidation test. J. Geotech. Geoenviron. Eng. **140**(6), 04014020 (2014). https://doi.org/10.1061/(asce)gt.1943-5606.0001103

Gorman, C.T., Hopkins, T.C., Dean, R.C., Drnevich, V.P.: Constant rate of strain and controlled gradient consolidation testing. Geotech. Test. J. **1**(1), 3–15 (1978)

Hamilton, J.J., Crawford, C.B.: Improved determination of preconsolidation pressure of a sensitive clay. In: Special Technical Publication No. 254, pp. 254–270. American Society for Testing and Materials, Philadelphia (1959)

Lee, K., Chao, V., Lee, S.H., Queek, S.H.: Constant rate of strain consolidation of Singapore marine clay. Geotechnique **43**(3), 471–488 (1993)

Sheahan, T.C., Watters, P.J.: Experimental verification of CRS consolidation theory. J. Geotech. Geoenviron. Eng. **123**(5), 430–437 (1997). https://doi.org/10.1061/(asce)1090-0241(1997)123:5(430)

Smith, R.E., Wahls, H.E.: Consolidation under constant rate of strain. J. Soil Mech. Found. Div. ASCE **95**(2), 519–539 (1969)

Wissa, A.E.Z., Christian, J.T., Davis, E.H., Heiberg, S.: Consolidation testing at constant rate of strain. J. Soil Mech. Found. Div. ASCE **97**(10), 1393–1413 (1971)

Yoshikuni, H., Moriwaki, T., Ikegami, S., Xo, T.: Direct determination of permeability of clay from constant rate of strain consolidation tests. Proc. Int. Symp. Compress. Consol. Clayey Soils **1**, 609–614 (1995)

Experimental Study on the Effects of Emulsion Wax Coating Agent (EWCA) About Isolation Performance of Cement Concrete Pavement's Transitional Layer

Jianbo Yuan[1], Huicong Wang[2(✉)], Jialiang Yao[2,3], and Yi Lin[2]

[1] Changsha University of Science and Technology, Changsha, Hunan 410114, China
yuanjb01@163.com
[2] School of Traffic and Transportation Engineering, Changsha University of Science and Technology, Changsha, Hunan 410114, China
{770692963,812782481}@qq.com, yao26402@126.com
[3] Key Laboratory of Road Structure and Material of Ministry of Transport (Changsha), Changsha, Hunan 410114, China

Abstract. The emulsion wax coating agent is a kind of new material about the isolation layer of cement concrete pavement (CCP). It could formed isolation layer between surface layer and base layer (also called emulsion wax isolation seal coat) when it sprayed on the base course surface of CCP. The CCP's emulsion wax isolation layer was provided with functions of isolation, friction reduction, waterproofing and so on, which could reduced the degree of bonding between the surface layer and the base layer, to weak the formation of weak transition layer (abbreviated "transition layer") in construction process of the concrete pavement's surface. The research of CCP's emulsion wax isolation layer is still remained in the "qualitative" level at present. With electronic microscope, mercury porosimeter and so on. This paper developed the research of set the structure of EWCA's isolation layer pavement (macrostructure, mesoscopic and micro structure) and the infiltration of wax isolation agent. And further through the surface of base course texture depth test, infiltration test, anti-erosion test, interlayer adhesion and friction coefficient test and other methods, bond strength between layers and friction coefficient test which analysis in-depth the research of wax isolation layer's mechanism of action in aspect of isolation, reduce friction, watertightness, improved the performance of the transition layer and so on, which established the coulomb model of interlayer adhesion and friction, etc. And indicated the mechanism of action of CCP's emulsion wax isolation layer.

1 Introduction

Because of base surface was uneven and the basis material of base was a gap when building cement concrete pavement, cement mortar easily penetrated the grassroots when panel was cast, thereby generated a transition layer between base layer and surface layer. The bonding of surface course and base layer will separating and

D. Chen et al. (eds.), *Advancements on Sustainable Civil Infrastructures*,
Sustainable Civil Infrastructures, https://doi.org/10.1007/978-3-319-96241-2_7

producing relative displacement under the action of temperature and load. The presence of the transition layer will cause part voids between surface layer and base layer which will result panel is non-evenly braced, base layer is broken, cracked and loss integrality, the bottom of the panel and the top of the base layer are exist different defects which resulting in panel or grassroots cracked. The cracking probability improves with the increasing of friction resistance of grassroots due to the temperature deformation of the surface layer according to the report of US Federal Highway Administration (FHWA), cracking of CCP which caused by air and temperature shrinkage produced tensile stress at large frictional resistance generally occurs during construction or hardening of cement concrete. This cracking which due to the bonding of the surface layer and base layer may begin at the bottom of panel and then to the panel, or may be reflecting to the surface layer because of shrinkage and cracking of base layer. In this regard, the project often lays a middle layer between the grass-roots level and the surface course, or calls isolation layer, the main function of the isolation layer is to reduce the level of bonding with the grass-roots level, and to eliminate or reduce weak transition layer which caused by the lower surface of concrete surface layer under the surface of the construction process. There are three main ways to set the isolation layer: [8] the first is paving asphalt materials which is used in emulsified bitumen slurry seal or paving a certain thickness (3 ~ 6 cm) asphalt mixture for the isolation layer in CCP's structural design theory and method of home and aboard. The second is lays of thin layer materials, includes laying geotextiles, plastic film, linoleum, etc.; the third is sprays emulsion materials which forms isolation layer after curing. Comparison of the three methods, geotextile, plastic film, linoleum as the isolation layer which easily affected by its upper construction equipment and appears moving, curling and damaging the surface layer, affecting the transition layer performance and isolation; When the slurry seal is too thin and its shear resistance is small. It easily forms weak interlayer when there is interaction of dynamic water between layers. The effect of interlayer of asphalt concrete paving more than 3 cm is good, but the cost is high, and because the bonding of asphalt and CCP will make the transition layer still exists and the friction resistance of interlayer is still high. The emulsion materials is non-toxic and tasteless, construct fast, low cost, good permeability, the representative material is emulsified paraffin and it forms the isolation layer of which have good friction and viscosity reduction performance.

This paper combined with the Guangdong Province qinglian highway project and so on carried out experimental study of isolation layer of EWCA. It developed that microstructure research of set wax isolation layer pavement's transition layer by MRI machine, further explored the effect of EWCA to performance of concrete pavement interlayer by different spaying amount of EWCA and different texture depth of base course surface test and adhesion of lower layer and friction coefficient test. The paper in-depth analyzed research of wax isolation layer's mechanism of action in aspect of isolation, reduce friction, watertightness, improved the performance of the transition layer and so on, established the coulomb model of interlayer adhesion and friction etc. And preliminary indicated the mechanism of action of CCP's emulsion wax isolation layer.

2 EWCA Materials and Requirements of Isolation Layer Construction

The option of EWCA is mainly in accordanced with the curing and isolation function requirements of EWCA, this refers to the relevant domestic and foreign standards, literature and the relevant test that carried out in this article to determine. The time of film-forming is refers to the time splashing water can form shape of oil droplets after spaying EWCA to surface, it generally need 1 ~ 5 h, it can be used to determine the minimum interval time between spraying EWCA to paved surface and the time of spraying that does not cover the plastic film before raining. For the using of ground concrete pavement isolation, the using of solid content not less than 25%, EWCA's ratio of high molecular material and mass ratio of paraffin is 1:2, the spraying dosage is 0.4 L/m^2, the minimum of spraying dosage is not less than 0.3 L/m^2 when used to isolation of filed concrete pavement. Ground construction uses special machinery spraying to ensure uniformity and controls the amount of spraying.

3 Experimental Study on the Effects of EWCA to Microscopic Structure of Transitional Layer

Transition layer refers to the undersurface layer of cement concrete surface layer, which is a feeble layer in the jointed reinforced concrete pavement (JPCP) (is shown in Fig. 1). Transition layer's interface shear and adhesive strength are lower than the surface layer strength of concrete pavement. Its internal gaps are more, the surface

Fig. 1. Isolation layer of concrete pavement

cracks are more. So there are more defects than panel. Under the bottom of panel attach isolation layer material, the panel easily happens stress concentration in the action of external load in plate, so that resulting in cracking of panel because of the exit of surface cracking of isolation layer. Due to the uneven of separation surface, panel easily happens fatigue failure under the effect of traffic loads, and then effects strained condition and reliability of road. The performance of the transition layer is mainly related to the construction of the concrete panel, the performance of concrete and the type of isolation adopted in the road.

This paper took core in ground (is shown in Fig. 2) by spaying EWCA of concrete pavement and no spaying EWCA of concrete pavement, the bottom of isolation layer

Fig. 2. Core sample figure of concrete pavement

of concrete sample's bulk density, porosity, pore distribution and permeability was tested by NMR machine and carried out imaging experiment and explored micro-capabilities of isolation layer concrete. NMR has advantages that samples will not be damaged and convenient and efficient as an emerging new method of core experiment analysis. In theory, pore fluid signal of rock samples is directly observed by NMR which not only can reveal pore structure characteristics of rock, but also it is possible to obtain the physical performance parameters of rock and well provides the method of multi-parameter analysis and comparison for the same piece of sample.

This article's took core of concrete strength grade is the design value of bending tensile strength 5 MPa, base course is C20 lean concrete base, the using of NMR machine is produced by Newmai Analytical Instruments Co., Ltd., MesoMR23-60-H small size MagModule II, its resonant frequency is 23.4028 MHz, the strength magnet is 0.5 T, the diameter of coil is 60 mm, the temperature magnet is 32 °C. The primary test results are as follows.

3.1 Volume Density

Part concrete sample of transition layer were cut from core sample to determine the bulk density of the transition layer. The results were shown in Table 1. The results of Table 1 were shown that the concrete volume density of the concrete pavement panel

Table 1. The volume density of transition layer

Number of samples	Dry weight (g)	Saturated water weight (g)	Volume (cm^3)	Volume density (g/cm^3)
No EWCA	106.0900	108.4764	44.52	2.38
Have EWCA	88.5800	86.3766	36.43	2.43

with no spraying EWCA was 2.38 g/cm^3 and the concrete volume density of concrete pavement panel with spraying wax curing agent was 2.43 g/cm^3. The results of test were shown that transition layer concrete is dense after spraying EWCA and its volume density was higher than the volume density of transition layer concrete with no spraying EWCA.

3.2 Porosity

Using a set of standard samples that known porosity to calibrate and established the relationship of porosity and signal. Signal collected by samples which is taken into the formula of porosity to calculate the porosity of sample. The results of this paper are as follows: The concrete porosity of concrete pavement panel transition layer with no spraying EWCA was 8.20% and the concrete porosity of concrete pavement panel transition layer with spraying EWCA was 4.50%. The results of test showed that the porosity of transition layer concrete is reduced, and beneficial to concrete dense after spraying EWCA.

3.3 Pore Diameter Distribution

The T2 spectra of samples (transverse relaxation time spectrum of pore fluid) was collected by using NMR analysis software. The pore size distribution of sample was calculated. The results of this paper are shown in Figs. 3 and 4.

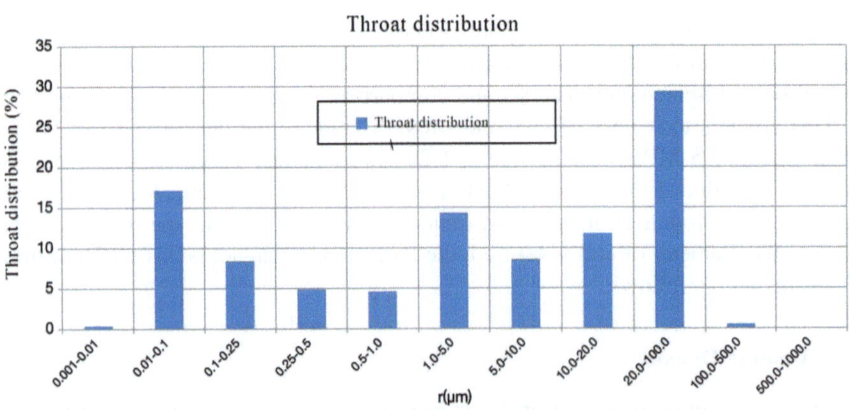

Fig. 3. The pore radius distribution histogram of transition layer with no EWCA isolation

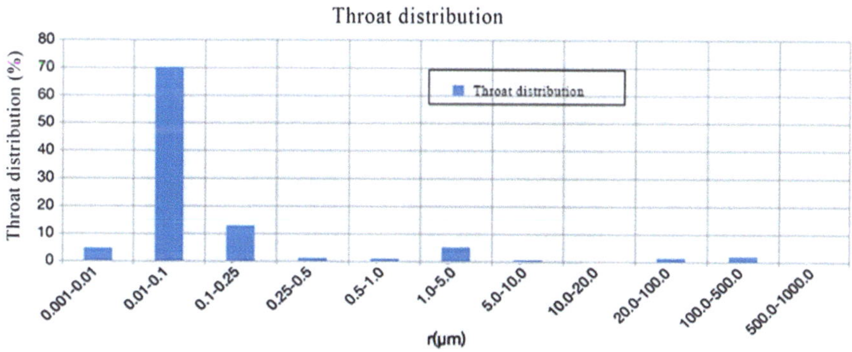

Fig. 4. The pore radius distribution histogram of transition layer with EWCA isolation

The results of test was shown that the main peak of pore radius of transition layer is 0.01 ~ 0.1 μm when EWCA was used. The main peak of pore radius of transition layer is 10–100 μm when EWCA wasn't used, which showed that the pore structure of concrete transition layer is improved, the number of small pores increased and the number of large pores decreased when EWCA was used.

3.4 Permeability

T2 spectra cutoff values was calculated after centrifugating. The permeability was calculated according to the Coates model. The results of this paper are as follows: the permeability of concrete pavement panel transition layer with no spraying EWCA was 34.9779 (mD), the permeability of concrete pavement panel transition layer with spraying EWCA was 0.0096 (mD). The results of test showed that the permeability of transition layer concrete is reduced and beneficial to concrete dense after spraying EWCA and beneficial to concrete dense after spraying EWCA.

3.5 Imaging Experiments

Proton density's weighted image of samples was collected and was shown in Fig. 5. It can be seen that transition layer concrete samples of concrete pavement panel without spraying EWCA has poor imaging effect and the relatively bright part of image is larger from Fig. 5. (The relatively bright part of image represents that the part of unit volume of sample contains more water).

4 Experimental Study on the Effect of EWCA on the Interlayer Performance of Concrete Pavement

This paper designed roughness++, roughness+, roughness, comparatively roughness and comparatively skin rolling in these five different kinds of texture depth of cement stabilized crushed stone base's sample surface by adjusting the grading of base

Number	Grayscale	pcolor	Chroma strip
Concrete transition layer with spraying EWCA	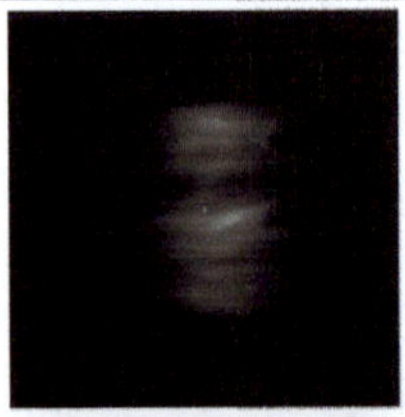	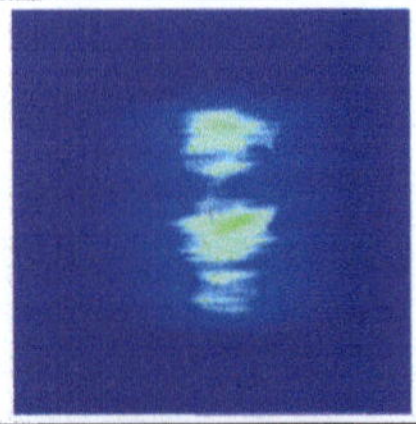	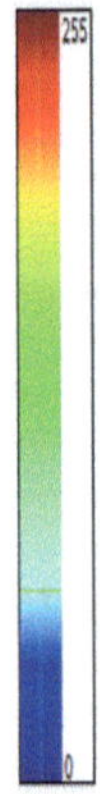
Concrete transition layer with no spraying EWCA		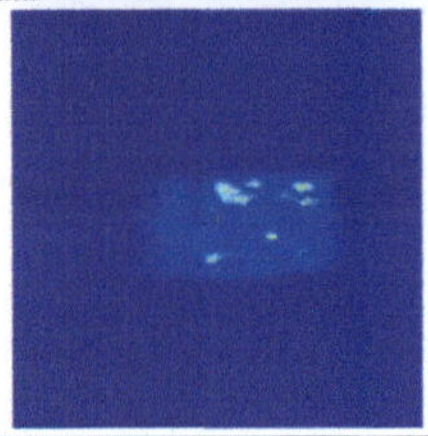	

Fig. 5. Density of Proton's weighted image

materials and after physically handling. The texture depth was measured by manual sanding method. The five kinds of different roughness of cement stabilized crushed stone base's texture depth were shown in Table 2.

Table 2. Texture depth of cement stabilized crushed stone base

The degree of roughness	Roughness ++	Roughness +	Roughness	Comparatively roughness	Comparatively skin rolling
Texture depth (mm)	2.14	1.74	1.18	0.77	0.39

Cement stabilized crushed stone base of different roughness degree were sprayed different dosage of EWCA, it was carried out pendulum friction coefficient test after 3d which is shown in Fig. 6. The BPN variation of different texture depth base surface is shown in Fig. 7.

The results of test is shown that:

(1) Different roughness degree of cement stabilized crushed stone base's BPN have been significantly reduced after spraying EWCA.
(2) Although at the beginning of spraying EWCA of dosage (spraying 0.2 L/m^2) is very small, reduction of BPN is also very obvious. Such as based on pendulum value was tested with no spraying EWCA, different roughness degree of cement stabilized crushed stone surface layer were sprayed small dosage of EWCA after transiting which was tested reduction extent of pendulum value was maximum.

Fig. 6. Figure of pendulum friction coefficient test

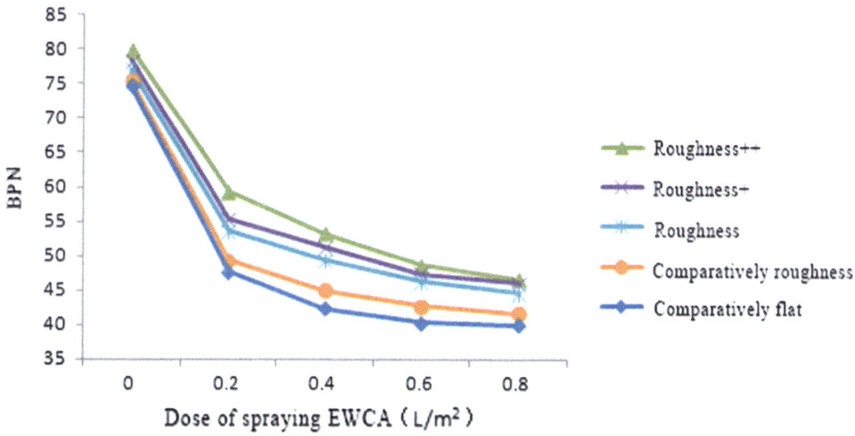

Fig. 7. Variation figure of BPN

Such as roughness++, roughness+, roughness, roughness and relatively flat of cement stabilized crushed stone with spraying a small dose of 0.2 L/m^2, the pendulum value respectively decreased by 36, 35, 30, 29, and 26%.

(3) The reduction of pendulum value has a tendency of slowdown with increasing of spraying dosage. Such as the amount of spraying increased to 0.4 L/m^2, based on pendulum value was tested when the spraying dosage is 0.2 L/m^2, roughness++, roughness+, roughness, relatively roughness and relatively flat of cement stabilized crushed stone of pendulum values relative to base were reduced by 7, 5, 6, 5, and 7%.

Through strength is C35 and 150 × 150 × 550 mm of trabeculae in the field molding, trabeculae's weight is 29.7 kg, the comprehensive friction coefficient test which is exerted a horizontal thrust to the end of samples and tested the maximum level of thrust when it was broken after compressive strength reached 30–35 MPa (The strength requirements of drawing test are the same). To simulate shear stress and frictional force that may produce when the concrete panel is constrained during the deformation process.

Comprehensive friction coefficient test uses heavily loaded truck to provide counter force, which is refers let car jack center and the center of proving ring arranges in the same level and centring. Recording the process that using car jack slowly pushing until trabecular specimen completely separated from the base layer when proving ring's loading achieving maximum. Maximum inter-laminar shear stress is determined by the ratio of maximum horizontal thrust $T_{\max}$ to shear area A ($150 \times 550\ \text{mm}^2$).

$$\tau = T_{\max}/A \tag{1}$$

Comprehensive friction coefficient f_h is the ratio of maximum jacking force of trabeculae $T_{\max}$ to the weight of trabeculae m.

$$f_h = T_{\max}/m \tag{2}$$

The variation trend of its comprehensive friction coefficient is shown in Fig. 8.

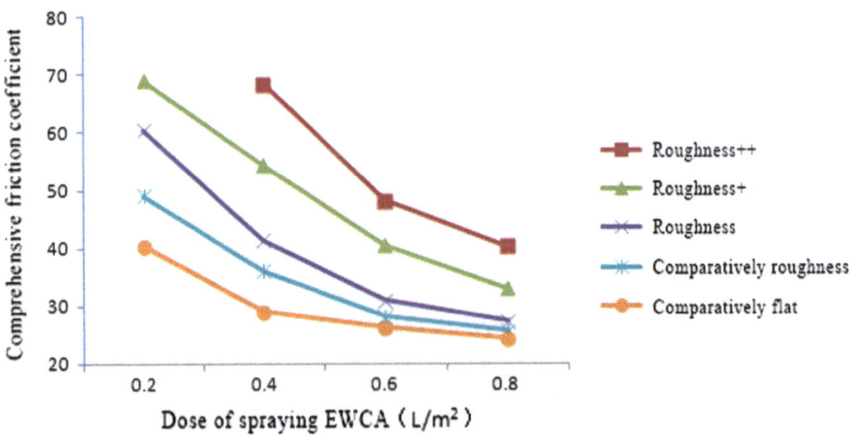

Fig. 8. Figure of comprehensive friction coefficient

The results of test is shown that:

(1) Cement stabilized crushed stone base's interlayer bonding was weaken after spraying EWCA on different roughness degree of cement stabilized crushed stone base surface. Low-dose (spray $0.2\ \text{L/m}^2$) sprays EWCA can significantly reduce the interlayer cohesion and comprehensive friction coefficient.
(2) The reduction extent of interlayer bonding stress and comprehensive friction coefficient decreasing with increasing of amount of spraying EWCA.

(3) According to the comprehensive friction coefficient can predicts, comprehensive friction coefficient will be constantly in a certain range when dosage of spraying EWCA reaches a certain amount, increasing the spraying amount of wax curing agent to the effect of comprehensive friction coefficient will no longer be so obvious, and will tend to a constant.

4.1 Different Base Surface of Mohr-Coulomb Formula

Adhesion performance of concrete pavement surface course can be used to determine its adhesion stress and frictional resistance by direct shear test in the indoors and outdoors, it can be described by Mohr-Coulomb theory:

$$\tau = C + \sigma \tan \phi \tag{3}$$

In the formula: τ is shear stress of interface, σ is normal stress, C is bond force of base layer and surface layer, $\tan\phi$ is internal friction angle of interface.

From the formula (3) can be seen that shear stress consist of two parts: adhesion stress and frictional resistance. Adhesion stress reflects the size of the interlaminar bond which is produced by the cementation of compound in the concrete and numerically equal to shear stress when the vertical stress is zero in the τ–σ curve of direct shear test. $\tan \phi$ is internal friction angle of interface which generally values $\tan \phi = f = 1.5 \sim 2.0$ in the common concrete pavement design and continuously-used. Internal friction angle ϕ reflects the effects of friction and interlocking between particles.

According to the maximum shear stress be measured under different normal stress conditions, the τ–σ curve of cement concrete and base layer interface can be made under different interlayer conditions. Direct stress test in this paper, the normal stress which applied to the surface of different isolation layer was separately 248, 496 and 744 N according to three standards, the pressure area is 150 mm × 550 mm, the corresponding normal pressure stresses is 3006, 6012 and 9018 Pa. Horizontal thrust was shown in Table 3, the specimen is prism, its size of 150 mm × 150 mm × 550 mm,

Table 3. Horizontal thrust of shear test

Type of isolation layer	Horizontal thrust P1 (N)	Horizontal thrust P2 (N)	Horizontal thrust P3 (N)
Base surface with no curing agent	226.4	462	577.5
Base surface with spraying EWCA	169.4	300.3	454.3

shear area is 150 mm × 550 mm. The existing tests show that τ–σ curve is not a strict straight line, but when the case of normal stress is not big (σ < 10 MPa), it can be approximated as a straight line. According to the straight line of τ-intercept to obtain cohesion of interface C, the internal friction angle of interface is determined according to the slope of straight line, base surface's texture depth of this paper was measured (sand patch method) not more than 0.3 mm, locality not more than 0.6 mm. The results of direct shear test on different surface layers are shown in Table 4.

Table 4. The results of shear test

Type of isolation layer	$\tan\phi$	Internal friction angle ϕ (°)	Bonding force C (MPa)
Base surface with no curing agent	0.7079	35.2947	8.6×10^{-4}
Base surface with spraying EWCA	0.5745	29.8774	2.8×10^{-4}

The results of test is shown that: the same texture depth of cement stabilized crushed stone base layer's interlayer bonding weakened and internal friction angle and bonding force weakened in coulomb equation after spraying EWCA.

5 Conclusions

(1) The test results of NMR is shown that the corresponding concrete pavement panel isolation layer's volume density was increased after base surface be sprayed EWCA. Porosity, pore size distribution and permeability of isolation layer concrete were improved.
(2) Different roughness degree of cement stabilized crushed stone base layer's BPN and synthetic friction coefficient have been significantly reduced after spraying EWCA, and reduction was slowing down with the increasing of spraying dosage.
(3) The same texture depth of cement stabilized crushed stone base layer's interlayer bonding was weakened and internal friction angle and bonding force was weakened in coulomb equation after spraying EWCA.

Acknowledgements. This research was supported by the National Natural Science Foundation of China under Grant 51578080, Grant 51178064 and the Transportation Department of Hunan Province, No. 2013-01.

References

Coates, G.R., Xiao, L., Prammer, M.G.: NMR Logging: Principles and Applications [M]. Halliburton Energy Services, Michigan (1999)

Li, T., Guo, H., Li, H., et al.: Research on movable fluids in shale gas reservoirs with NMR technology [J]. Spec. Oil Gas Reserv. **19**(1), 107–109 (2012)

Morriss, C., Rossini, D., Straley, C., et al.: Core analysis by low-field NMR [J]. Log Anal. **38**(2), 84–94 (1997)

Okamoto, P.A., Naussbaum, P.J., Smith, K.D., Darter, M.I., Wilson, T.P., Wu, C.L., Tayabji, S. D.: FHWA-RD-91-079 Guidelines for Timing Contraction Joint Sawing and Earliest Loading for Concrete Pavements. Volume 1: Final Report [S]. FHWA, U.S. Department of Transportation (1994)

Voigt, G.V.: Early Cracking of Concrete Pavement—Causes and Repairs [DB/OL]. American Concrete Pavement Association (2002)

Wang, W.: The Physical Study of Nuclear Magnetic Resonance in Rock and Its Application on Petroleum Industry [D]. Institute of Physics and Mathematics (Chinese Academy of Sciences), Wuhan (2001)

Wang, X., Xiao, L., Xie, R., et al.: Study of NMR porosity for terrestrial formation in China [J]. Sci. China, Ser. G **49**(3), 313–320 (2006)

Wu, G.: Study on cement concrete pavement cracking mechanism and damaged process. Southwest Jiaotong University Doctor Degree Dissertation (2003) 9. (in Chinese)

Wu, G., Yi, Z., He, Z.: Study on cracking process of cement concrete pavement. Highway **10**, 141–143 (2010). (in Chinese)

Xiao, L., Shi, H.: Low field NMR core analysis and its applications in log analysis [J]. WLT **22** (1), 42–49 (1998)

Yao, J., Yuan, J., Zhang, Q.: Experimental study of emulsified wax curing agent and asphalt slurry seal as bond breaker media in cement concrete pavement [J]. China Civil Eng. J. **10**, 127–131 (2009)

Yi, Z., Wu, G., Yang, Q.-G., Wu, Z.-L., et al.: Failure process analysis of cement concrete pavement based on fracture mechanics and a new idea of pavement design [J]. J. Jiaotong Univ. **20**(1), 1–4 (2001). (in Chinese)

Yi, Z.-J., Tang, B.-M., Li, Z.-W., et al.: Basic failure types of cement concrete 5 pavement caused by interaction between the surface course and the base course. J. Jiaotong Univ. **20**(1), 1–4 (2001). (in Chinese)

Yi, Z., Tang, B., Yang, Q.-G.: New structures of cement concrete pavements with bond breaker mediums [C]. The Corpus from the 1st National Forum of Highway Scientific and Technical Innovation, 84–88 (2003). (in Chinese)

Evaluating Settlement on King Faisal Road Using GIS Analysis Techniques

Mahmoud Abdelrahim Abdelgiom[1,2] and Muhammad Mubaraki[1](✉)

[1] Department of Civil Engineering, Faculty of Engineering, Jazan University, Jazan, KSA, Saudi Arabia
mmubaraki@jazanu.edu.sa

[2] Faculty of Engineering, Department of Surveying Engineering, University of Khartoum, Khartoum, Sudan

Abstract. This study aims to determine the reasons for settlement on King Faisal Road in Jazan City, Kingdom of Saudi Arabia, using Geographical Information System (GIS) techniques. Pavement engineers and contractors in Saudi Arabia are often faced with the challenges of constructing pavement for desert sand dune areas, mountainous areas, and the sabkha or salt flat areas, and they are faced with problems of settlement. King Faisal Road in Jazan City is selected as a case study. This road suffers distress problems such as settlement and bleeding. The primary objective of this paper is to investigate the reasons for the settlement on King Faisal Road in Jazan City by using two criteria of investigations. These criteria of investigation include soil and asphalt laboratory test results, and analysis of the site's drainage system by Geographic Information System (GIS). The obtained results show that the reasons for settlement along King Faisal Road were focused on the drainage systems on the road and the traffic loading characterized by different classes of vehicles. The purpose of this investigation was to determine the cause of settlement inconsistencies in pavement surface elevations at the distressed locations. There are different distressed locations on King Faisal Road, but the main distressed site is located in the area of intersection between King Faisal Road and Prince Sultan Road, encompassing the intersection's traffic signal domain. Proposed solutions to the problem of settlement are presented in this paper. One of these solutions: Drilling under the road layers (sub-base, base, and asphalt) to a depth of at least five meters and drying the area using the appropriate means of water drainage such as vertical drilling wells. The second proposed solution: Using a chemically-enhanced solution to a depth of at least ten meters under the road layers, with a soil injection of high-softness cement, to increase the durability of the soil in the settlement areas.

Keywords: King faisal road · GIS · Settlement · Jazan city · ArcGIS10.1 Archydro10.1

D. Chen et al. (eds.), *Advancements on Sustainable Civil Infrastructures*,
Sustainable Civil Infrastructures, https://doi.org/10.1007/978-3-319-96241-2_8

1 Introduction

The network of roads in Saudi Arabia has experienced remarkable development, both in total length and in standards of quality, to become one of the longest and most modern highway networks in the Middle East. The total length of paved roads was 12,200 km in 1975, but currently the length is in excess of 40,000 km. The government has invested more than $30 billion in road construction since 1975.

Pavement engineers and contractors in Saudi Arabia are faced with the challenges of constructing pavement for desert sand dune areas, mountainous areas, and in the sabkha or salt flat areas, and they are faced with problems of settlement in these different terrains. Each area has its own difficulties and requirements in terms of design, specifications, and construction practices. At the same time, the quality of materials varies widely, according to their availability in each area [1].

This study is carried out with the problems of settlement (see Fig. 2) along King Faisal Road and focused on the drainage systems on the road and the traffic loading characterized by different classes of vehicles. The purpose of this investigation was to determine the cause of settlement in pavement surface elevations. The study is applied in the different distressed locations on King Faisal Road, and is focused on the main distressed site. This site is located in the area of intersection between King Faisal Road and Prince Sultan Road, encompassing the intersection's traffic signal domain.

A problem of settlement along this road is caused a large number of defects such as longitudinal cracks (see Fig. 2) or lane-to-lane staggering was observed in the pavement after the road construction [2].

2 Literature Review

2.1 Settlement in General

Settlement in road construction is usually caused by the weight of fill material, not by traffic. Traffic loads are transient. Traffic loads can cause additional compaction, but that influence is usually only in the upper 12–18 in.

Engineers need to consider a higher-modulus material as the fill, in order to "bridge" the softer soils. Unfortunately, higher modulus usually means greater compaction and higher unit weights. To counteract this, I have specified cellular concrete fill, which has a unit weight of about half that of soil, for such applications. I designed a truck haul road for a paper mill over a reclaimed swamp area in the year 1997. I also designed a high-modulus fill without reducing weight for a similar application, but with mostly automobile traffic, for a municipal street in 1992. Both lasted at least their anticipated design lives. I used a conventional flexible pavement section for both applications, above the high-modulus fill [3].

This activity is performed on roads where experience has shown that pavement is likely to be subject to excessive movement as a result of the road's location, its foundations, amount of traffic, and other factors which may now be different from those for which the road was designed. The need to monitor pavement may also develop with time, as factors such as climate change, water table variations, traffic density changes, soil compaction, and changes in soil foundation type, all take effect [4].

2.2 Settlement Limit and GIS

The total settlement is the magnitude of downward movement. Differential settlement is non-uniform settlement. It is the difference of settlement between various locations of the structure. Angular distortion between two points under a structure is equal, to the differential settlement between the points divided by the distance between them.

What can be considered tolerable, in settlements of different structures, varies considerably. Simple-span frames can take considerably greater distortion than rigid frames. A fixed-end arch would suffer greatly if the abutments settle or rotate. For road embankments, storage silos, and tanks, a settlement of 300–600 mm may be acceptable, but for machine foundations the settlement may be limited to 5–30 mm. Different types of construction materials can withstand different degrees of distortion. For example, sheet metal wall panels do not show distress as readily as brick masonry [5].

To reduce differential settlement (ΔH), the designer may limit the total settlement and use the following equation for the calculation of the differential settlement:

$$(\Delta \mathrm{Hdiff})\,\mathrm{max} = {}^{1}\!/_{2}\,\Delta \mathrm{Htotal} \tag{1}$$

Information regarding attributes of local human habitation and road use is important for decision makers at all levels of planning, as they have to grapple with dynamic environments often associated with instances of road settlement. At the local level, it is particularly important for both communities and urban managers have accurate and reliable information regarding all planning attributes. The use of Geographical Information System (henceforward, GIS) tools in settlement mapping and upgrading provides a platform for integration of spatial and non-spatial data necessary for planning and decision making. The capabilities offered by these tools in GIS projects have enabled communities to participate in settlement planning and upgrading, including the management of new infrastructure.

Roads are the main arteries of modern society's infrastructure, contributing heavily to the distribution of goods and persons. GIS provides many helpful applications for ensuring a smooth flow, by aiding design, routing, traffic control, and real-time navigation. In essence, a GIS application in transportation may no longer be a GIS, but instead a merger of GIS with Intelligent Transportation Systems or Transport telemetrics, where GIS no longer exists as a stand-alone product. This essay will attempt to display the extent of existing GIS applications in road transportation, and to critically assess their appropriateness and potential [6].

GIS provides a valuable tool in the process of planning and designing roads. This is related to the term Computer Aided Design (CAD), but it is hard to tell at what level of detail CAD stops and GIS actually begins. Modern software (e.g., Bentley's Micro-Station) tends to bridge this gap between discipline-specific applications and GIS in such a way that they are fully integrated. A GIS can help visualize and communicate the effects of roads on their environment. Engineering drawings and maps may evoke a vivid landscape in the minds of engineers familiar with them, but to decision makers or the public in general, these drawings can be quite incomprehensible. Traditionally, displaying different route options and proposals has been done in the form of 2D maps, assisted by section drawings, maybe together with an aerial photo, where the road network was overlaid in the form of lines. It is simple and straightforward, but it is not conveying much information on the actual impact of the pavement design. Therefore, the UK Highways Agency has taken on building virtual showrooms, presenting the road by displaying a 3D drive-through along a highway or a bird's-eye view of the landscape from any angle at which the user wants to hover [7]. This can be presented to a planning board or the public using ordinary PCs, or it can even be published on the Internet for open access. The highways agency in the French department Loire has been using GIS since 1989 for many purposes: Traffic accident patterns are visualized and safety improvements are made where they are most needed. By capturing significant data for the whole network, repairs and works budgeting have become more reliable and can be calculated in advance. First creating the optimal route between locations and then using GIS to decide how and where to position road signage, has improved traffic direction and movements in the road network, and has helped in avoiding congestion [8].

The construction of an embankment on the weak subsoil always leads to settlements. The magnitude of these settlements depends on the magnitude of the loading, i.e., the dead weight of the embankment, and on the deformation characteristics of the subsoil. As both the magnitude of the loading and the deformation characteristics of the subsoil usually exhibit some variation, the settlements of the subsoil nearly always are non-uniform. Nonuniform settlements result in unevenness of the road or railway, and therefore they have a negative effect on driving comfort and even on traffic safety [9].

3 Objectives

The primary objective of this paper is to describe the detailed investigation that has been conducted to study the causes of settlement along King Faisal Road in the city of Jazan, using soil and asphalt laboratory tests and GIS analysis techniques.

To achieve this objective, a field survey program was conducted in 2007 on King Faisal Road. Photographs of settlement areas (see Fig. 2) on the road surface were taken directly at the road sites during the field survey in order to evaluate existing conditions and related road obstacles noted by visual inspection.

This field survey was carried out in order to investigate the causes of the settlement at the study site. Soil and concrete-asphalt mixture specimens were taken directly from the road sites to illustrate existing conditions and related road obstacles, and to aid in our study of low levels of performance linked to areas of settlement in the road.

The secondary objectives are:

a. To investigate the reasons for low pavement performance on King Faisal Road using modern software facilities (ArcGIS software)
b. To design a successful pavement model for King Faisal Road, in order to use this model in Routine (Periodic) Maintenance
c. To investigate how GIS (2D, and 3D) improves spatial analysis

d. To evaluate and find the causes of pavement distress on King Faisal Road, and
e. To compare and discuss the results of soil and asphalt laboratory tests obtained and to extract valuable conclusions and recommendations for the site.

4 Methodology

The methodology of this paper is classified into two parts, a description of the study area and a detailed case study.

4.1 Study Area

King Faisal Road is one of the most important streets in Jazan City, connecting east and west through the central part of Jazan. At 2.8 km in length and 60 m in width, this road consists of dual carriageways and eight lanes, divided equally by a central island. Its median width is 6 m.

The study was concentrated on a critical part of the road, selected as a case study due to serious problems and failures caused by the sub-grade of expansive soil and the accumulation of subsurface water in the pavement structure. The road suffered from severe distresses of settlement, rutting, and heavy depressions. Rehabilitation of the road was carried out two times a year by the Alamana (Municipality) of Jazan City (see Figs. 1 and 2). In the study area, the importance data is captured are:

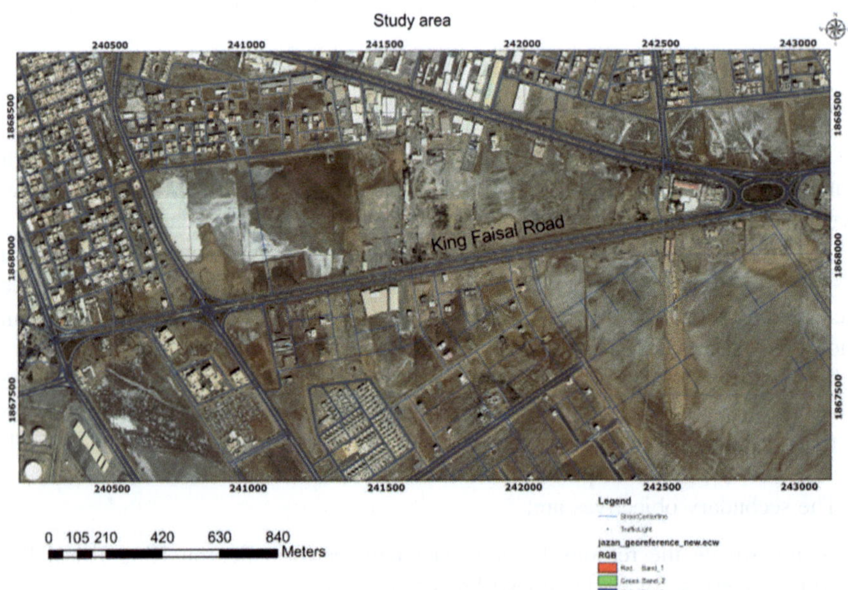

Fig. 1. Layout map of King Faisal Road (derived from ArcGIS software)

Fig. 2. Settlement areas in the study area

- Soil specimens in the settlement locations on the road
- GIS attributes (descriptive) data such as: (road name, road center line coordinates, road lengths, road widths, road lanes, road medians, Current road pavement condition, traffic lights, etc.)
- Photographs are taken directly from the road sites during the field survey.

4.2 Case Study

In the case study the following steps were followed to fulfill the objectives of the project.

a. Data including detailed surveying of natural ground levels and existing paved level of King Faisal Road, soil and asphalt specimens were taken from the specific settlement location in the road, attribute databases, and the most necessary GIS meta-data for the project were collected
b. A digital model of King Faisal Road using GIS facilities was established
c. Analysis using GIS software was performed
d. A computer programs to achieve the objectives of this research, mainly ArcGIS10.1, Archydro10.1, Excel, and Origin70.exe, were used and
e. The results of soil and asphalt laboratory tests, and suggested solutions were discussed to solve the settlement problem of the study site, King Faisal Road in Jazan City, Saudi Arabia.

5 Data Analysis

The analysis was carried out in this paper to investigate the reasons for the observed settlement. The investigations included:

a. Soil tests and standardization of results according to AASHTO classifications, and
b. Drainage systems checked by GIS software (Arc Hydro).

5.1 Analysis of Soil Specimens of King Faisal Road

Analyses of the soil and asphalt specimens selected from site survey in the settlement locations of the king Faisal Road were carried out in soil and asphalt laboratories located in Jazan City. The significant results of this testing included the following:

5.1.1 Moisture Density Relationship of Soil

The Moisture Density Relationship of Moisture Content and Dry Density of soil is graphed as Moisture Content (%) versus Dry Density (g/cm^3), shown in.

Table 1 shows the test data of bore hole specimen (Coordinate: X = 240,723, Y = 1,867,785). This data recorded in the laboratory is entered into the Moisture content table and the Final Moisture Content calculated.

Table 1. Moisture content (%) versus Dry density (g/cm^3) of King Faisal Road

Moisture content (%)	3.47	4.93	6.67	8.08	10.43
Dry density (g/cm^3)	1.911	2.067	2.178	2.195	2.145

The difference between the soil graph, Moisture Density Relationship of King Faisal Road (The left part), and the AASHTO graph (The right part) is shown in Fig. 3.

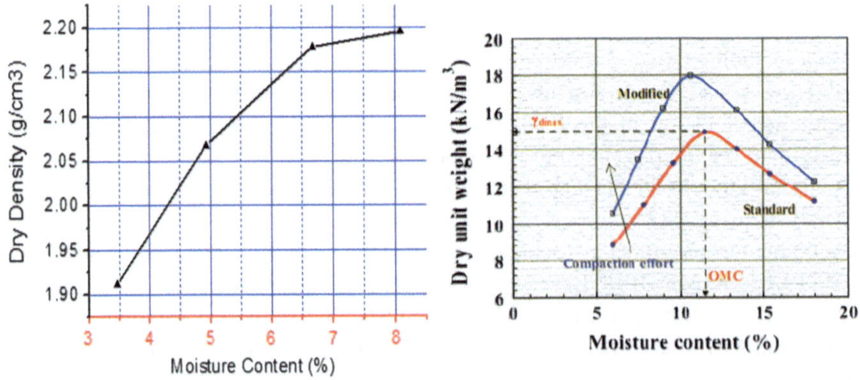

Fig. 3. Moisture content (%) versus Dry density (g/cm^3) of King Faisal Road and the AASHTO Standard Graph

5.1.2 Sieve Analysis Test

The result of sieve testing of specimens obtained at the King Faisal Road site (Coordinate: X = 240,723, Y = 1,867,785) is shown in Table 2.

Table 2. Sieve diameter versus passerby %

Sieve number	Sieve diameter (mm)	Passerby %
100	38.1	1.5″
92.53	25.4	1″
87.76	19	0.75″
75.52	12.5	0.5″
69.23	9.38	0.375″
56.71	4.75	#4
47.65	2.36	#8
41.22	2	#10
33.2	0.425	#40
29.25	0.355	#50
0.83	0.075	#200

In this test, Reserved percentage is calculated by Eq. 2:

$$\text{Reserved}\,\% = \left[\frac{W}{W_t}\right] * 100 \tag{2}$$

where: W_t= Total weight of mass (W_t) = 1446 g.

W = Reserved weight (g).

The comparison between Sieve diameter versus Passerby % of the King Faisal Road soil graph and the AASHTO graph is shown in Fig. 4.

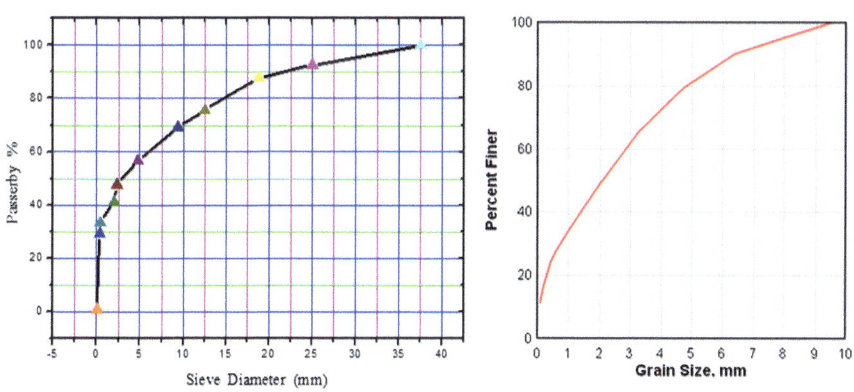

Fig. 4. Graph of sieve diameter versus passerby % and the AASHTO graph

5.1.3 California Bearing Ratio (CBR) Test

The CBR—value is used as an index of soil strength and bearing capacity. This value is broadly used and applied in the design of base and sub-base material for pavement. In this King Faisal Road study, the specimen of soil (Coordinate: X = 240,723, Y = 1,867,785), is taken from the base layer. The results of the analysis the bore hole specimens are shown in Tables 3 and 4.

Table 3. California bearing ratio (CBR) test (CBR% and No. of blows)

CBR % at 2.54 mm	26	46	65
CBR % at 5.08 mm	34	51	73
No. of blows	10	30	65

Table 4. The California bearing ratio (CBR) test (CBR% used and compaction)

Penetration (in.)	Stress (kg/cm^2) in 10 blows	Stress (kg/cm^2) in 30 blows	Stress (kg/cm^2) in 65 blows
0.05	3	6	8
0.1	5	9	12
0.2	8	12	17

The comparison between the California Bearing Ratio (CBR) test (Penetration vs. Stress) of the King Faisal Road soil graph and the AASHTO graph is shown in Fig. 5.

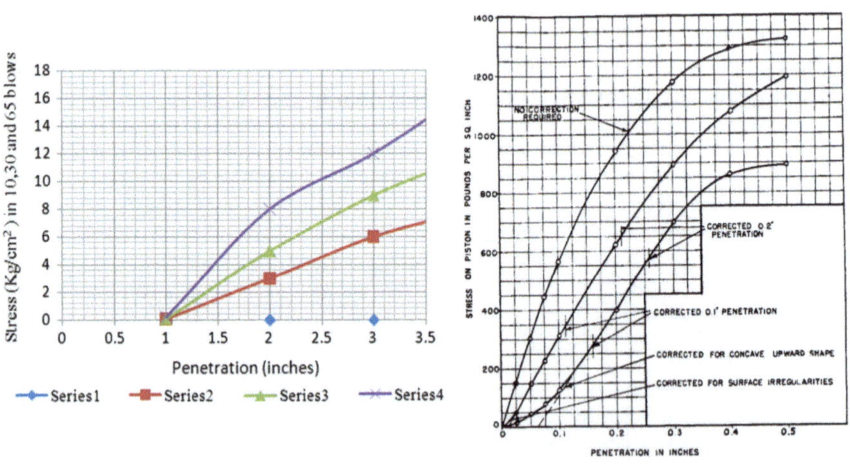

Fig. 5. Comparison of the CBR test graph and the AASHTO standard graph

5.1.4 Marshal Test

Calculations of the Marshal Test

According to the results of the Marshal Test in the bore hole (Coordinate X = 240,720, Y = 1,867,788), the calculations of the Marshal Test carried out are:

- Find bulk unit weight (γ) from Eq. 3:

$$\gamma\,(\mathrm{g/cm^3}) = \frac{W_a}{v} = \frac{W_a}{W_a - W_w} \tag{3}$$

where:

W_a: weight of specimen in air (g), W_w: weight of specimen in water (g) V: volume of specimen = volume of displaced water (cm^3).

- Determine % of air void by determining maximum theoretical unit weight from Eq. 4

$$\Psi\,(\mathrm{g/cm^3}) = \frac{W_a}{W_b \times G_b + W_c \times G_c + W_f \times G_g + W_{mf} \times G_{mf}} \tag{4}$$

Then determine % void in the total mix (i.e., in specimen) from Eq. 5:

$$\%\mathrm{V.T.M} = \frac{\Psi - \gamma}{\Psi} \tag{5}$$

where:

V_b, W_b, G_b = Volume (cm^3), Weight (g) and specific gravity (g/cm^3) of binder,

V_c, W_c, G_c = Volume (cm^3), Weight (g) and specific gravity (g/cm^3) of course aggregate,

V_f, W_f, G_f = Volume (cm^3), Weight (g) and specific gravity (g/cm^3) of fine aggregate,

V_{mf}, W_{mf}, G_{mf} = Volume (cm^3), Weight (g) and specific gravity (g/cm^3) of mineral filler, and

V.T.M. = Voids in the total mix.

The summarization of this test is shown in Table 5.

Table 5. Bitumen % by volume and Stability (correction) kg

Bitumen % by volume	3.920	3.91	3.900
Stability (correction) kg	2486	2116	2283

The difference between Bitumen by the volumes of the King Faisal Road soil graph and the AASHTO graph is shown in Fig. 6.

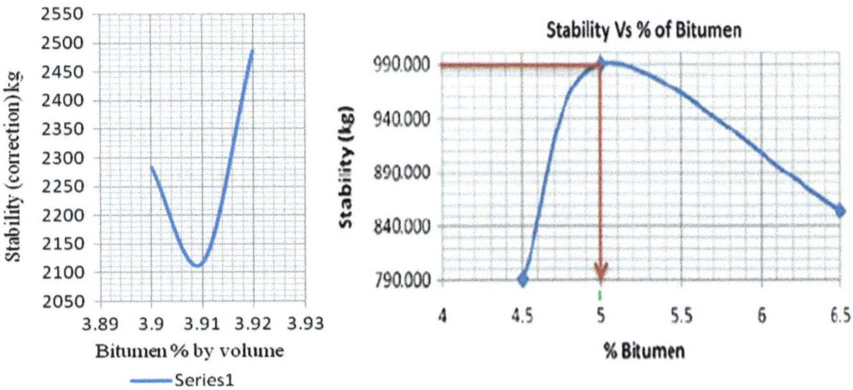

Fig. 6. Comparison between the Bitumen by volumes (%) graph and the AASHTO standard graph

5.2 Drainage Systems of Jazan City Using GIS

Arc Hydro is used mainly by hydrologists or GIS professionals interested/working in the field of water resources. Even if you are not directly interested in water resources, it is at least beneficial to know some basic processes conducted with a very powerful and extensive set of tools.

In this paper the GIS analysis carried out depending on the spot heights of Kazan City. These elevations within Jazan are taken according to height above mean sea level. The necessary steps within ArcHydro10.1 software were used to investigate the main drainage of rain surface water in Jazan City. The main steps of GIS analysis using ArcHydro10.1 include:

- Open ArcGIS 10.1, add Archydro tool, and Set Processing Output
- Reconditioning of interpolating data for Jazan City
- Fill Sinks of Digital Terrain Model (DTM) of Jazan City
- Flow Direction (see Fig. 7)
- Flow Accumulation
- Stream Definition
- Stream Segmentation
- Catchment Polygon Processing and Drainage Line Processing (see Fig. 8).

6 Discussion of Results

The investigations carried out at the King Faisal Road study site are summarized in the following discussion of the results:

- Figure 3 shows the clear difference between the Moisture Density Relationship of the King Faisal Road soil graph and the AASHTO graph. This showed that the percentage of Moisture Density was increased from time to time because there are

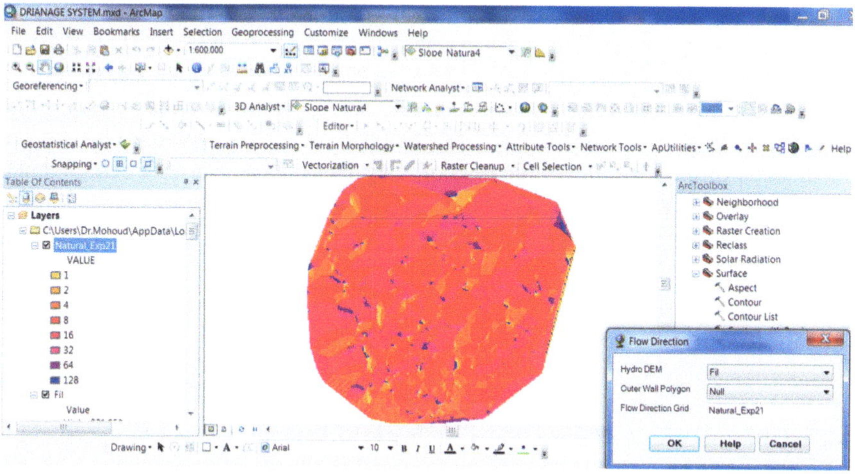

Fig. 7. Flow direction of Jazan City

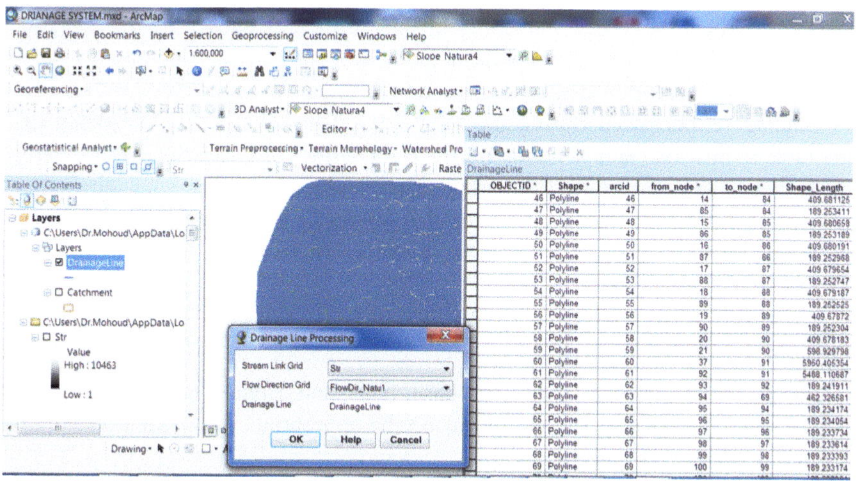

Fig. 8. Drainage line processing of Jazan City

areas of open space beside the road. These areas are very full of water during the autumn season. The depth of the water table in the study area is about 1.8 m. The facilities of GIS helped the researchers to investigate the drainage system easily. The drainage results are calculated in this project by use of the ArcHydro10.1 software. The performance of King Faisal Road would improve if these areas were rehabilitated by property owners or by the government of Jazan City.

- Figure 5 shows no difference between the Sieve diameter vs. Passerby % of the King Faisal Road soil graph and the AASHTO graph. This means that the aggregate of the sub-base layer of this road is selected according to AASHTO classifications.

- Figure 8 shows the difference between the California Bearing Ratio (CBR) test (Penetration vs. Stress) of the King Faisal Road soil graph and the AASHTO graph, this led to the poor bearing capacity and the high value of the expected decline.
- Figure 10 shows the difference between bitumen by volumes versus stability of the asphalt of the King Faisal Road soil graph and the AASHTO graph. This caused the settlement of the road; the traffic load of King Faisal Road causes a certain amount of damage to the pavement structure. This gradually accumulates over time and eventually leads to pavement failure.

7 Conclusion

The effort made in this research project clearly revealed the potentiality of applying modern GIS software (ArcGIS10.1 and Archydro10.1) to the analysis of the conditions for King Faisal Road. In this test, the type of soil in the study area is classified as the Silty sand (SM) Soil. The depth of bore hole equal 25 cm.

The settlement founded in the King Faisal Road pavement is very adverse, and it decreases the sustained period of the road. It causes pavement distresses and deterioration which affect the safety and riding quality of the pavement.

Our analyses showed that the percentage of Moisture Density increased from time to time because there are open spaces beside the road. Problems increased due to the poor drainage experienced during the rainy season in Jazan City. The depth of the water table in this road area is about 1.8 m. The facilities of GIS helped the researchers to investigate the drainage system easily. The results of drainage analysis were calculated in this paper by using Archydro10.1 software. The performance of the study area would improve if these areas were rehabilitated by property owners or by the government of Jazan City.

Finally, proposed solutions to the problem of settlement are presented in this paper include the following:

(a) Drilling under the road layers (sub-base, base, and asphalt) to a depth of at least five meters and drying the ground using the appropriate means of water drainage (such as vertical drilling wells).
(b) Using a chemically-enhanced soil solution to a depth of at least ten meters under the road layers, with a soil injection of high-softness cement, to increase the durability of the soil in the settlement areas.
(c) Good soil should be applied with a thickness of not less than 2 m. The replacement is carried out by placing a layer of auxiliaries conforming to the American specifications (AASHTO) and being of type(A-1-a) − (A-1-b) The layer for (20 cm) should achieve a greater grading grade than (97%) than the modified Proctor density for the same soils. This method also requires the prevention of access of water to the soil as well as the isolation of reinforced concrete and ensures that groundwater does not reach it.

(d) Reduce the usability of compression in the study area by pressure classes under road layers by mechanical means such as compaction.
(e) Confirmations of soil tests should be made to determine the effectiveness of the treatment methods applied.

References

1. Bubshait, A.A.: Quality of pavement construction in Saudia Arabia. Pract. Periodical Struct. Des. Constr. **6**(3) (2001)
2. Chen, Y., Lu, Y., Zhang, Z.: The cause of longitudinal pavement fissures and preventions on highway road-widening engineering. East China Highway, 2003, China
3. www.eng-tips.com/viewthread.cfm
4. www.mainroads.wa.gov.au/BuildingRoads
5. Skempton, A.W., MacDonald, D.H.: The allowable settlement of buildings. London (1956)
6. Husdal, J.: Road transportation management using GIS—vehicle routing and tracking. Unpublished course paper for the M.Sc. in GIS. University of Leicester, UK (1999)
7. Leslie, S.: The importance of being geographic: why geographic information analysis is critical to the modern organization, UK (1999)
8. Marshall, J.: Point de depart: GIS as a driving force in the Loire, GIS Europe **4**(6) 1995
9. Pilot, M., Moreau, M.: Calculation method for the stability of embankments on weak sub-soils. France (1973)

Effect of Using Sludge Waste Containing Calcium Carbonate on the Properties of Asphalt and Hot Mix Asphalt Mixture

Varinder Kanwar[1(✉)], Er. Abhishek Kanoungo[1], and Sanjay Kumar Shukla[1,2]

[1] Chitkara University, Himachal Pradesh, India
{registrar,abhishek.kanoungo}@chitkarauniversity.edu.in
[2] Discipline of Civil and Environmental Engineering, Edith Cowan University, Perth, Australia

Abstract. Fatigue failure is one of the common forms of distress in asphalt pavements and manifests itself in the form of cracking under repeated traffic loading or a series of temperature fluctuations/variation in the pavement. It is caused by gradual buildup of irrecoverable strains under repeated loading, which develop into a measurable rut (permanent depression along the wheel path). These strains are due to the visco-elastic response of bituminous materials to dynamic loading. Unmodified bitumen lacks the balance between the visco-elastic response and the increase of the traffic volume. Therefore, binders were modified to face the load and weather challenges. Modified bitumen introduces advantages to the field of highway construction, by improving pavement performance as well as extending the pavement life. Recently, numerous waste materials have resulted from manufacturing operations, service industries and household in which several millions of non-biodegradables are produced. If not taken care of they will persist in the same state for a considerable time and thus cause hazardous problems to environment and human life. Therefore, in the present work, an attempt is to use sludge containing calcium carbonate ($CaCO_3$), which is a waste product from industry as an asphalt modifier. This will contribute to the utilization and effective disposal of waste sludge. The effect of using lime modified asphalt in bituminous mix is the main thrust of this research work.

Keywords: Waste lime sludge · Modified bitumen mix · Conventional test

1 Introduction

An efficient road network is necessary for the national growth, industrial and socio-economic development of the country. The purpose of highway pavement is to provide smooth surface over which vehicles can move safely from one place to another. The vehicular density has increased over the last few years due to improvement in living standard of the community. Likewise, the damage to roads have also increased because of the continuous wheel load being transferred to the pavement, sometimes

D. Chen et al. (eds.), *Advancements on Sustainable Civil Infrastructures*,
Sustainable Civil Infrastructures, https://doi.org/10.1007/978-3-319-96241-2_9

exceeding the bearing capacity, which results in defects like rutting, fatigue, low skid resistance causing the failure before its design life. Thus, the concern for the maintenance of highways and major roads has risen amongst the government and national authorities. Engineers seek to hold these forms of failure to acceptable limits within a pavement design life.

The ability of the asphalt mixture to withstand repeated bending without fracture is termed as fatigue resistance. It is one of the common forms of distress in asphalt pavements and manifests itself in the form of cracking under repeated traffic loading or a series of temperature fluctuations/variation in the pavement causing deformation. Permanent deformation is caused by gradual buildup of irrecoverable strains under repeated loading, which develop into a measurable rut (permanent depression along the wheel path). These strains are due to the visco-elastic response of bituminous materials to dynamic loading. Figure 1 [1] shows the visco-elastic response to millions of wheel loadings.

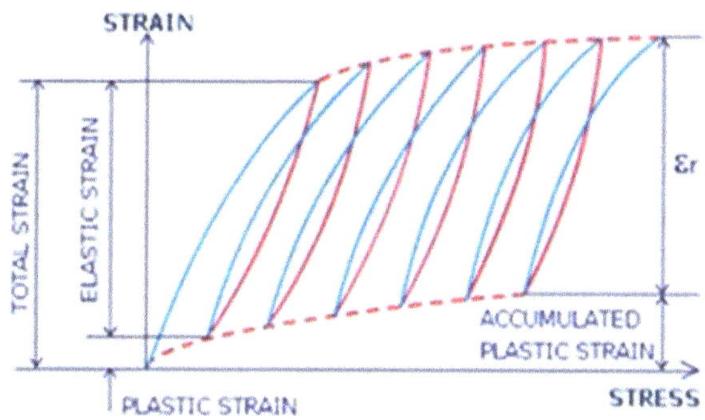

Fig. 1. Accumulated Plastic strains in pavements [1]

Unmodified bitumen lacks the balance between the visco-elastic response and the increase of the traffic volume. In addition, over the life of a pavement unmodified bitumen has limited capacity at wide range of loads and temperature. Therefore, binders were modified to face the load and weather challenges. Modified bitumen introduces advantages to the field of highway construction, by improving pavement performance as well as extending the pavement life.

Therefore, in the present work, an attempt is made to use sludge containing calcium carbonate ($CaCO_{3,}$) which is a waste product from industry as asphalt modifier. This will also contribute to the utilization and effective disposal of waste sludge. Different tests for the prepared asphalt $CaCO_3$ blend will be conducted. These include the test for both physical and the mechanical properties of asphalt and mixtures without and with $CaCO_3$. The effect of using lime modified asphalt in bituminous mix is the main thrust of this research work.

2 Literature Review

Palit [2] discussed that Crumb rubber modified mixes displayed lower potential for permanent deformation compared to normal mixes. 30CR10 mixes showed the least potential for permanent deformation. Crumb rubber modified mixtures were found to be less susceptible to moisture damage compared to normal mixtures as indicated by higher retained Marshall Stability, higher tensile strength ratio, and improved stripping characteristics.

Mohammad [3] experimentally proved that, the appropriate amount of polyethylene was determined to be (6–18%) by weight of the optimum asphalt percent (5.4%), which equates to (0.34–1.03%) by weight of total aggregate. However, the optimum modifier content was found to be 12%, which equals only 0.68% of the total aggregate weight. Overall, using the polyethylene in asphalt mixture reduces pavement deformation; increase fatigue resistance and provide better adhesion between the asphalt and the aggregate.

Kok and Yilmaz (2009) studied the effects of SBS and lime as mineral filler in hot mix asphalt. They found that the stability of unconditioned lime treated mixtures was approximately 8% higher than those of the unconditioned control mixture. However, this value increased up to 21% for the conditioned mixtures. According to retained Marshall Stability they concluded that the addition of only 2% lime had approximately same effect with addition of 6% SBS about moisture damage.

Gorkem and Sengoz [4] aimed to determine the effect of additives such as hydrated lime as well as elastomeric (SBS) modified bitumen on the stripping potential and moisture susceptibility characteristics of hot mix asphalt (HMA) containing different types of aggregate (basalt–limestone aggregate mixture and limestone aggregate). The results indicated that hydrated lime addition and polymer modification increased the resistance of asphalt mixtures to the detrimental effect of water.

Iwnaski and Mazurek [5] discussed on the basis of the rheological properties, that the recovered binder with hydrated lime had a positive effect on reducing aging process of bitumen 50/70 modified with SBS polymer in SMA mixture giving the increase in resistance to the water and frost effect.

Behiry [6] concluded that both lime and cement could increase Marshall stability, resilient modulus, tensile strength and resistance to moisture damage of mixtures especially at higher condition periods. Use of hydrated lime had better results than Portland cement.

Cui et al. [7] studies the effect of water on the adhesion of bitumen to three aggregates (limestone, marble and granite). Acidic aggregates (granite) showed a greater loss of adhesion than basic aggregates (limestone and marble) under wet conditions. The porosity of the aggregates, although shown to be significant, was less important than their chemical composition. The interfacial adhesion in wet conditions can be improved by mixing a saline, amine or rubbery polymer into the bitumen.

Abril et al. [8] studied the behavior of bituminous mixes made with sewage sludge ash as mineral filler. He concluded that SSA waste may be used as filler for bituminous mixes with better results than for mixes made with limestone fillers and with similar results for mixes made with other fillers such as hydrated lime and cement.

Ogundipe [9] indicated that the replacement of the mineral filler with lime improves the stability of the mixtures, while there was slight increase in the flow of the mixture with hydrated lime. The slight increase in stability and flow values may be attributed to the complete replacement of the mineral filler with lime and the high lime content used in the study.

3 Materials and Methods

3.1 Material Selection

This study focuses on addition of lime sludge that is generated in bulk quantity from toothpaste and paper industry to bitumen, thereby improving the properties of bitumen. The utilization will also reduce the menace caused due to non-disposal of sludge in an effective manner.

- **Lime sludge**
  Lime sludge was collected from toothpaste and paper industry. Before reusing any waste material, it is necessary to characterize the material. The chemical characteristics of lime sludge are given in Table 1.

Table 1. Chemical characteristics of lime sludge

Test carried out	Test value
Loss on ignition% by mass (@900° ± 25°)	63.5
Silica as SiO_2 % by mass	0.18
Iron as Fe_2O_3 % by mass	0.04
Alumina as Al_2O_3 % by mass	0.36
Magnesium as MgO % by mass	0.93
Calcium as CaO % by mass	34.2
Sodium as Na_2O % by mass	0.11
Titanium as TiO_2 % by mass	0.0006
Potassium as K_2O % by mass	0.07

Other characteristics: The various other characteristics are mentioned below:
Reactivity test: The sample was tested with respect to its reactivity with water. It does not evolve fumes, explosive gases etc. on addition of water. Hence, the sample does not indicate reactive nature.
Ignitability test: On ignition, the sample does not show catching fire.
Corrosively test: The pH of the water sample is found to be 7.3. Hence the sample does not indicate the chance of corrosive nature (the acceptable range is 2–12.5).

- **Bitumen**
  Bitumen is a viscous liquid consisting of hydrocarbons which is soluble in tri-chloro-ethylene and softens gradually when heated. In our research, VG-10 grade bitumen was used. The properties of unmodified bitumen used are given in Table 2.

Table 2. Properties of unmodified bitumen (VG-10 grade)

Penetration value (1/10th of mm)	Ductility, cm @ 27 °C	Viscosity (s)	Softening point (°C)	Marshall stability (kg)	Flow value (mm)
95.3	78.7	43	51	1601.4	3.21

- **Aggregates**
  Locally available coarse aggregates were used. Coarse aggregate for bituminous mix has been defined as that portion of the mixture which is retained on 2.36 mm. Aggregate passing through 2.36 mm sieve and retained on 0.075 mm sieve was selected as fine aggregate. Tests were performed to determine the Aggregate Crushing Value, Aggregate Impact Value, Specific gravity, L.A. Abrasion value and Water absorption and results are summarized in Table 3. The specific gravity of coarse aggregates is 2.7 and that of fine aggregates is 2.76 Fine sand and stone dust finer than 0.075 mm sieve were used as filler in bituminous mixes having specific gravity as 2.48.

Table 3. Properties of aggregates used in the study

Properties	Coarse aggregates	Fine aggregates
Aggregate crushing value (%)	22	–
Impact value (%)	18	–
Specific gravity	2.7	2.76
LA abrasion value (Grade C)	17	–

3.2 Preparation of Modified Bitumen

The bitumen was heated to fluid condition at a temperature of 120 °C after which the powdered sludge containing calcium carbonate was added. The blend was then manually mixed for about 5–10 min. The modified bitumen was cooled to room temperature and then different tests were carried out.

3.3 Tests on Bitumen

The following tests were carried out on bitumen:

- Penetration test (IS: 1203–1978)
  It is used to determine the consistency of bituminous material. The test is performed using an apparatus called as penetrometer.
- Ductility test (IS: 1208–1978)
  Ductility of a bituminous material is its ability to stretch. It is measured by the distance in centimeters to which it will elongate before breaking.
- Softening point test (IS: 1205–1978)

The softening point is the temperature at which the substance attains degree of softening under special condition of test. It is determined by ring and ball apparatus.

- Viscosity test (IS: 1206–1978)
  The viscosity affects the ability of the binder to spread, move into and fill up the voids. It is determined as the time in seconds required by the material to flow from a cup through a specified orifice.

3.4 Marshall Stability and Flow Value Test

The Marshall stability of the bituminous mix specimen is defined as the maximum load carried in kg at the standard temperature of 60 °C when load is the applied. The flow value is the total deformation of Marshall Test specimen at the maximum load, expressed in mm units. It is used for determining the optimum bitumen content. The aggregates weighing 1200 g were taken of the desired gradation. Bitumen is heated to a temperature of 121–125 °C adding different percentages of sludge. The first trial percentage of bitumen say 4.5% by weight of the mineral aggregates is taken. The heated aggregates and bitumen are thoroughly mixed at a temperature of 154–160 °C.

4 Results

- **Penetration test values**—The results of penetration test are tabulated in Table 4 for varying percentages of sludge. Here we can see that as the percentage of waste increases, the penetration value of the mix decreases up to 20% of waste sludge after which it increases. Hence it can be said that addition of sludge converts soft grade into hard grade.
- **Ductility test values**—The results of ductility test are given in Table 5 for varying percentages of sludge. It shows the varying trend of ductility value with increasing percentages of sludge. As the percentage of sludge increases, the ductility value of the mix decreases up to 20% of waste sludge after which it increases. Therefore, the stretching ability of the bituminous mix decreases.

Table 4. Results of penetration test

Sludge added (%)	10	15	20	25
Penetrometer dial reading	71.66	52	35.33	41
Grade of bitumen	60/70	50/60	30/40	40/50

Table 5. Results of ductility test

Sludge added (%)	10	15	20	25
Mean ductility value (cm)	57.9	43	35.8	39.1

- **Softening point test values**—The results of softening point test are tabulated in Table 6 for varying percentages of sludge. As the percentage of sludge increases, the softening point of the mix increases up to 20% of waste sludge after which it decreases. Here it is observed that the heating temperature of bituminous mix increases on addition of sludge.

Table 6. Results of softening point test

Sludge added (%)	10	15	20	25
Softening point (°C)	62	65	69	66

- **Viscosity test values**—The results of viscosity test are tabulated in Table 7 for varying percentages of sludge. As the percentage of sludge increases, the viscosity of the mix increases. We know that at higher viscosity binder may not fill up the voids completely thereby resulting in poor density of the mix. Hence it should be verified that the viscosity value is under limit while adding sludge in the bituminous mix before it can be used.

Table 7. Results of viscosity test

Sludge added (%)	10	15	20	25
Viscosity (s)	52.5	61	68	64

- **Marshall Stability and Flow Value test**—The results of Marshall Stability and flow value are tabulated for varying percentages of sludge in Table 8. It is evident that the optimum binder content is achieved at 5.5% with addition of 20% sludge as the maximum strength is achieved at this value as 1988.3 kg.

Table 8. Results of Marshall stability and Flow value test

Sludge added (%)	10	15	20	25
Marshall stability (kg)	1459.4	1553.4	1988.3	1754.5
Flow value (cm)	2.98	3.43	2.40	2.89
Voids in total mix (%)	4.01	3.60	4.02	3.85
Voids filled with bitumen (%)	79.97	92.45	82.68	84.6
Optimum bitumen content (%)	5.5	6.5	5.5	5

- **Comparison between modified and unmodified bitumen mix for different components of Marshall Stability test are shown in Graphs 1, 2, 3 and 4.**

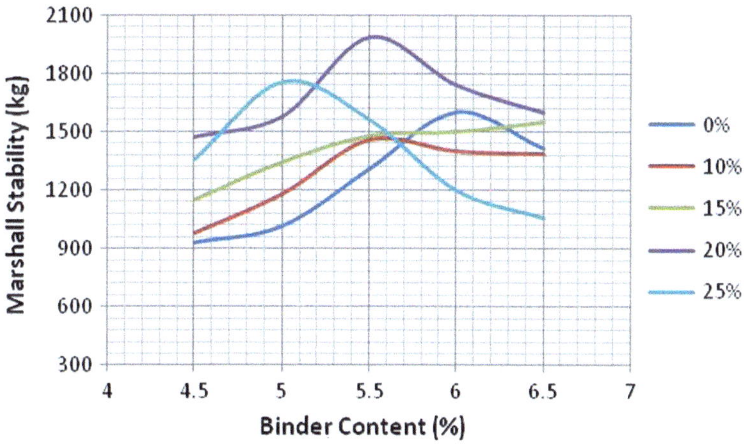

Graph 1. Relation between Marshall stability and Binder content

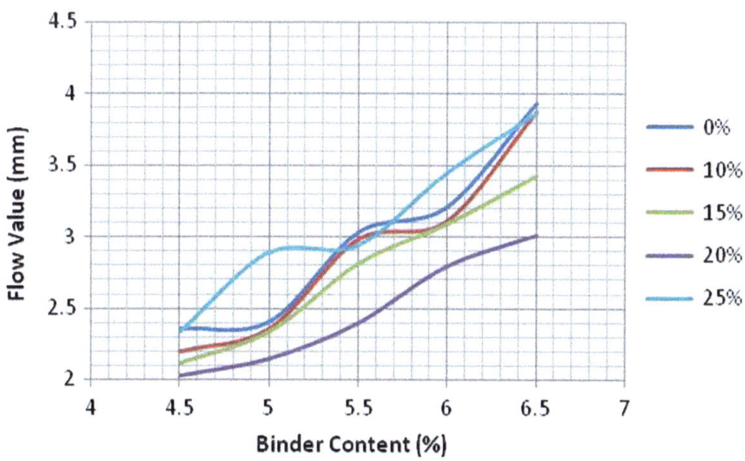

Graph 2. Relation between Flow value and Binder content

5 Conclusions

- After completing the conventional tests, it is concluded that soft grade bitumen can be converted into hard grade by proper mixing it with lime sludge.
- The optimum percentage of sludge that can be added to the bituminous mix is found to be 20% of total weight of the bitumen.

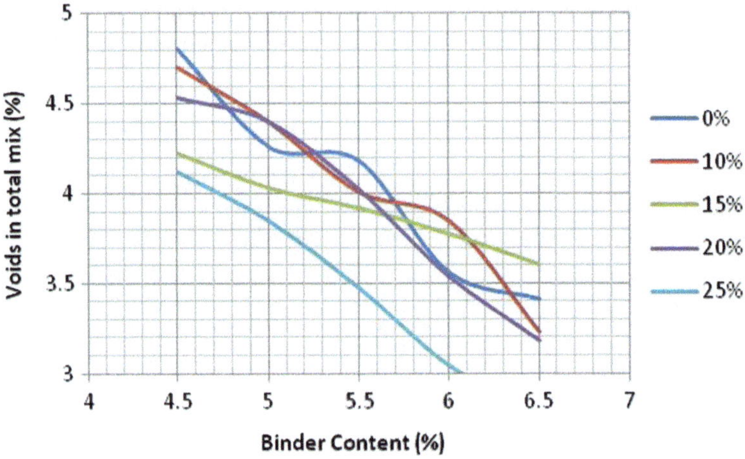

Graph 3. Relation between Voids in total mix and Binder content

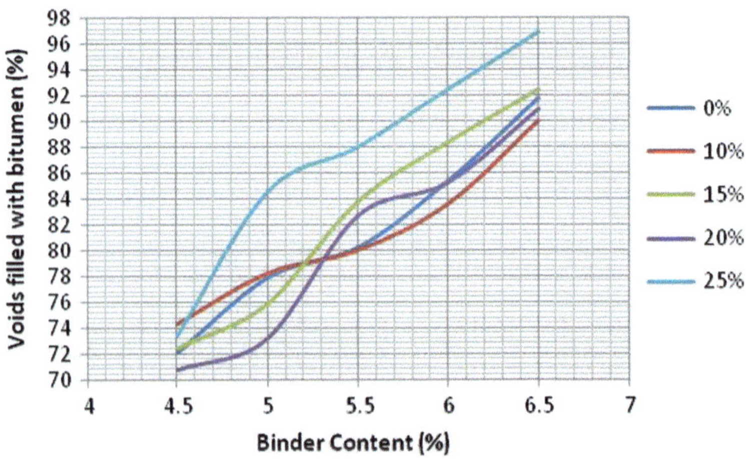

Graph 4. Relation between Voids filled with bitumen and Binder content

- According to ductility and viscosity test, the test results of the different mixes had increased value of viscosity and decreased ductility values in comparison to unmodified bitumen.
- The penetration value of the modified bitumen mix is found to be 35.33 which prove that it is hard grade bitumen. (Standard value 30–40).
- The Marshall Stability result of unmodified bitumen was 1601.4 for strength and 3.21 for flow value with OBC as 5.5%. The change in the value is observed with addition of waste sludge. The strength increases 24% with the addition of 20% of waste sludge and OBC as 5.5% and flow value decreases up to 26%. It is also seen

that as the percentage of sludge is increased, the strength tends to decrease. The use of lime will have a long time effect on stability by reducing ageing of the binder.
- Using waste sludge, we will eliminate the problem and consequences of disposing it in the environment. With some modification in design mixes, it can result in effective utilization of waste bituminous pavement, thus saving considerable investment in construction and partially solving the disposal of wastes.
- For success of such an initiative, the proposed material(s) should be safe, environmental friendly and cost effective. This paper concludes that we can use waste sludge containing $CaCO_3$ in road construction. By adding optimum percentage of lime sludge the soft grade bitumen i.e. 80/100 used in the present study was modified to hard grade bitumen i.e. 30/40.
- The paper identified that the test conducted are insufficient to examine the fatigue of the bituminous mixture. Future studies are being carried out to check the fatigue behaviour and deformation resistance of the bituminous mix.

References

1. Asphalt Institute.: Super pave TM Mix Design. Super pave Series SP-2, Lexington, Kentucky, USA (1996)
2. Palit, S.K. et. al.: Laboratory evaluation of crumb rubber modified asphalt mixes. J. Mater. Civil Eng. (2004). © ASCE/January/February 2004 https://doi.org/10.1061/asce/0899-1561/2004/16:1/45
3. Mohammad, T.: The use of polythene in hot asphalt mixtures. Jordan in Am. J. Appl. Sci. 390–396 (2007). ISSN 1546-9239 Science Publications
4. Gorkem, C., Sengoz, B.: Predicting stripping and moisture induced damage of asphalt concrete prepared with polymer modified bitumen and hydrated lime. Constr. Build. Mater. **23**(6) (2009)
5. Iwański, M., Mazurek, G.: Hydrated lime as the anti-aging bitumen agent. Proc. Eng. **57**, 424–432 (2013).
6. Behiry, A.E.A.E.M.: Laboratory evaluation of resistance to moisture damage in asphalt mixtures. Ain Shams Eng. J. **4**(3), 351–363 (2013)
7. Cui, S., Blackman, B.R. K., Kinloch, A.J., Taylor, A.C.: Durability of asphalt mixtures: effect of aggregate type and adhesion promoters. Int. J. Adhes. Adhes. **54**, 100–111 (2014) October 2014
8. Tenza-Abril, A.J., Saval, J.M., Cuenca, A.: Using sewage-sludge ash as filler in bituminous mixes. J. Mater. Civil Eng. **27**(4) (2015)
9. Ogundipe, O.M.: Marshall stability and flow of lime-modified asphalt concrete. Transp. Res. Proc. **14**, 685–693 (2016)
10. Yilmaz, M. et. al.: The effects of using lime and styrene-butadiene-styrene on moisture sensitivity resistance of hot mix asphalt. Constr. Build. Mater. **23**(6) (2009)
11. Penetration of bitumen BSEN 1426:2007 & ASTM D5–05a
12. Ductility of bitumen ASTM D113–99
13. Softening point-ring and ball BSEN 1427:2007 & ASTM D36–95 (2000)
14. Determination of viscosity of bitumen BSEN 13302:2010 & AASHTO T316-04 & ASTM D-4402-02
15. American Association of State Highway Transportation Officials [Aashto]

Mixture Design of Emulsified Asphalt Cold Recycled Mixture by Accurate Volume Method

Jinjin Shi[1(✉)], Yingbiao Wu[1,2], Chenfang Yang[1], and Xiuxian Li[1]

[1] Cangzhou Municipal Engineering Company Limited, Cangzhou, China
czszjsk@126.com

[2] Hebei Province Road Materials and Technology Engineering Technology Research Center, Cangzhou, China

Abstract. Plant mixed emulsified asphalt cold recycled mixture (EACRM) is a kind of new type of road material which can be used for low-grade road surface and high-grade road base. However, there is no universally accepted mix design method of emulsified asphalt cold reclaimed mixture. This paper discusses a new method of fitting design—accurate volume method. In this method, it uses rotary compaction molding of the mixture, the amount of emulsified asphalt is estimated by the regression formula, which is more representative than the 4% method in the Marshall design method. The design method also uses the effective compaction theory to evaluate the workability of the emulsified asphalt cold recycled mixture, and the optimum liquid dosage and the mixing water quantity are closer to the actual situation of the construction site. The performance of the mixture is verified, which proves that the method is a good design method.

1 Introduction

Since 1990s, many roads in China need maintenance and rehabilitation. During the period, a huge amount of reclaimed asphalt pavement (RAP) has been produced. Asphalt pavement cold recycling is an environmental protection technology which can use RAP materials extensively. This not only saves resources and protects the environment, but also saves 40–50% of the total investment when compared with other traditional way [1, 2].

According to the different types of stabilizers, asphalt pavement cold recycling can be divided into emulsified asphalt recycling, cement recycling, foamed asphalt recycling. Among them, emulsified asphalt cold recycling technology uses emulsified asphalt as a stabilizer to produce the cold recycled mixture, which can save resources, reduce environmental pollution and improve the old road surface performance, improve the old road level. Thus, it has a wide range of applications prospect.

However, at present, there is not an universal design method of emulsified asphalt cold recycled mixture (EACRM) [3]. In this paper, the traditional mix design method is discussed, and a new method of mix design is proposed, which is based on the accurate volumetric method of rotary compaction. The performance of the mixture is verified.

D. Chen et al. (eds.), *Advancements on Sustainable Civil Infrastructures*,
Sustainable Civil Infrastructures, https://doi.org/10.1007/978-3-319-96241-2_10

2 Design Method of Accurate Volume Method Based on Rotational Compaction

At present, most of emulsified asphalt cold recycled mixture design employs modified Marshall design method in China [4]. It is based on the experience of estimating the amount of emulsified asphalt and determination of the optimum moisture content through the heavy compaction after selection of raw materials, new mineral aggregates and emulsified asphalt. After Marshall Compaction, the optimum emulsified bitumen of the Marshall specimen was determined by the theoretical maximum density, the integrated relative density, the stability at 60 °C and the flow value.

In the practical engineering application, it is found that the optimal water content determined by the modified Marshall design method is quite different from the actual construction. Since the Marshall design method can't simulate the actual compaction effect of the pavement [5], the method can't be well simulate the actual compaction work in the construction site. The accurate volume design method based on the rotary compaction is to evaluate the optimum workability of the mixture and determine the best liquid content via the effective compaction work. The compacted samples via the cold recycling method are closer to the field compacted materials [6].

For the accurate volume method proposed in this paper, raw materials were tested and prepared. The aggregate mixture proportion was determined through design of the proportions of different types of aggregate. Then, the equation method was used to estimate the initial amount of emulsified asphalt; After the relevant parameters were prepared by pre-mixing and pre-compaction, the optimum total liquid dosage based on different water contents was determined based on the theory of effective compaction work. According to the void volume requirement of recycling mixture, the appropriate content of emulsified asphalt can be determined. Finally, the emulsified asphalt cold recycling mixture was tested through the spilt test.

3 Procedure of Accurate Volume Design

In order to explain the accurate volume design method of EACRM in detail, this work uses 60% old raw material emulsified asphalt cold recycled mixture as an example, the specific test steps include:

3.1 Selection of Raw Materials

RAP material came from in the Beijing–Shanghai Expressway Cangzhou section during large scale repair period. RAP material was piled in the storage yard after it was milled and crumbed. To reduce variances as far as possible in the sieving tests, some special ways have been used. For example, the RAP from upper, middle and bottom were batched as mass proportion of 1:1:1 and two different sizes RAP (0–5 mm and 10–30 mm) were used. Gradations of two sizes RAP was determined by the standard sieve test.

Emulsified asphalt in the test is Shell cationic slow crack emulsified asphalt, which has 66.8% evaporation residue (mass proportion).

The new aggregate is 10–25 mm limestone mineral material, whose test indexes met the specification requirements.

Cement in the test used Bohai 'Lion' brand PS32.5 Portland slag cement, and its performance test results met relevant requirements. The reason of adding a little cement in Emulsified asphalt cold recycled mixture is to improve emulsification, early stage strength and high temperature stability.

3.2 Equation Method to Estimate the Amount of Emulsified Asphalt

To design the combined gradation of aggregates, linear planning program method had been employed to calculate proportion of different aggregates. In the recycling mixture, the RAP was fixed to 60% (total aggregate mass). The combined aggregate gradation of the mixture is shown in Table 1, and its curve is shown in Fig. 1.

Table 1. The combined aggregate gradation of the mixture

Sieve size (mm)	Passing rate (%)					Combined gradation	Limits of the gradation [7]	
	RAP		New aggregate				Lower limit	Upper limit
	10–30	0–5	10–30	10–20	0–5			
31.5	100	100	100	100	100	100.0	100	100
26.5	93	100	84	100	100	96.7	80	100
19	76.4	100	13.6	85.5	100	82.9	–	–
16	65.3	100	4	60.5	100	76.0	–	–
13.2	43.3	100	1	33.5	100	67.5	60	80
9.5	0.3	100	0.2	4.5	100	54.7	–	–
4.75	0.1	58	0.1	0.1	96.7	36.9	25	60
2.36	0.1	30.4	0.1	0.1	73.1	23.5	15	45
1.18	0.1	16.4	0.1	0.1	47.4	15.3	–	–
0.6	0.1	9.4	0.1	0.1	33.3	11.1	–	–
0.3	0.1	5.2	0.1	0.1	23.7	8.5	3	20
0.15	0.1	2.1	0.1	0.1	12.9	6.2	–	–
0.075	0	1.0	0.1	0.1	4.35	4.9	1	7
Mixing ratio	20	40	12	14	10	Cement 2% Mineral Powder 2%	–	–

Each particle mixture included: 20% coarse milling material (RAP 10–30 mm), 40% fine milling material (RAP 0–5 mm), 12% coarse new aggregate (new size 10–30 mm), 14% medium aggregate(new size 10–20 mm), 10% fine aggregate (new size 0–5 mm), 2% Portland slag cement and 2% mineral powder.

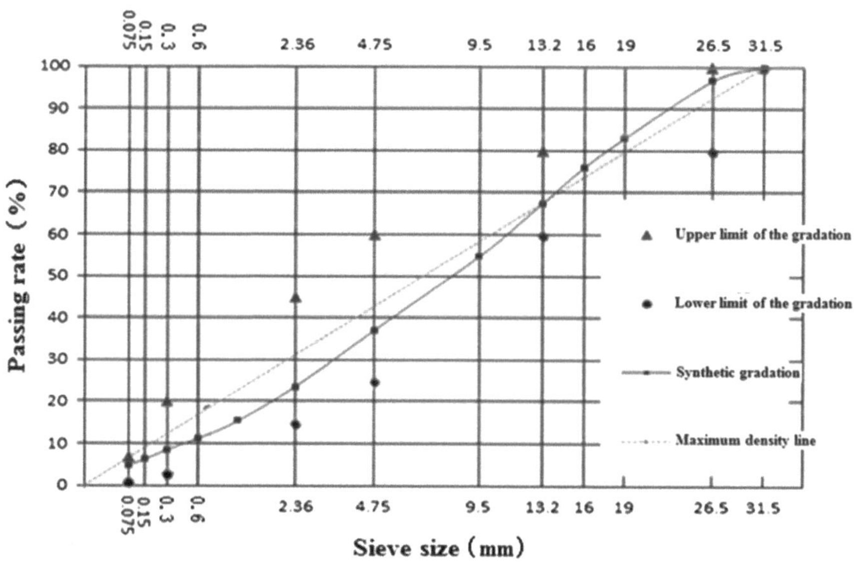

Fig. 1. Combined aggregate gradation curve

The content of emulsified asphalt emulsion followed below equation:

$$EA = \frac{(0.05B + 0.01C) \times 100}{A} \quad (1)$$

where EA is the estimated proportion of emulsified asphalt (the mass ratio of mixture); B is the percentage of aggregate smaller than 4.75 mm, which is 36.9; C = 100 − B, which equals to 63.1; A is the percentage of evaporation residue of emulsified asphalt, which is 66.8.

Equation (1) has a good relationship with the passing percentage of 4.75 mm sieve. According upper data, EA is 3.7% through calculation.

3.3 Relevant Parameters Based on Pre-mixing and Pre-compaction

The first step is to prepare new aggregate and RAP mixture according to the particle materials mass percentage. The total liquid is the total mass of emulsified asphalt and water, and its value ranges from 6 to 8% of the mixture. The mixture was mixed for 0.5–1 min in the automatic mixer with no liquid; then the water was added and mixed for another 0.5–1 min; and the emulsified asphalt was added to mix for 1 min. with emulsified asphalt; The mixture was kept in room condition at least 2.5 h after mixing uniformity.

In this test, a total weight of particle mixture material was 4500 g. The total liquid include 3.3% of water and 3.7% of emulsified asphalt based on the particle mixture mass. Specimen was compacted by shear gyratory compactor. Compaction pressure parameter was set to 600 kPa, the number of rotation compaction was 35 gyrations and the specimen height was 116.2 mm. There was a small amount of water squeezed from the bottom of the trial mold. Based on test experience, the total liquid 7% was a little more than the actual need.

3.4 Optimum Total Liquid Amount Based on Effective Work

The estimated value of the total liquid minus the estimated amount of the emulsified asphalt was the amount of water used for mixing. The estimated amount of the emulsified asphalt was kept constant, and the estimated value of the water consumption is the median value (abbrev. m-v), with a interval of ±0.5%. Mix with 7 water contents (m-v, m-v ±0.5%, m-v ±1.0%, m-v ±1.5%) for the compacted specimen by shear gyration compaction. Compaction curve can be calculated based on the effective work value.

According to the theory of effective compaction work, gyration compaction pressure of indenter and area of specimen were fixed. Therefore, the force on the loose mixture was a fixed value, so the work of per millimeter displacement was unchanged. For different mixtures with the same mass, the work on the specimen with maximum displacement was maximum, which means the best compaction effect. Compressive pressure of indenter set to 600 kPa, the effective compaction work calculation as follows:

$$\overline{W} = 10.603 \times \overline{\Delta H} \tag{2}$$

where $\overline{W}$—the effective work mean value J; $\overline{\Delta H}$—weighted mean of the displacement mm; $\overline{\Delta H} = \frac{h_i - h_j}{j - i}$ h_i—the corresponding specimen height mm at i times; h_j—the corresponding specimen height mm at j times; i, j-compaction times.

Therefore, effective work theory is used to calculate the effective work of the mixture with different water content. Through the total liquid content—effective work fitting curve, the peak of the curve is the best dosage of the total liquid.

This test was carried out under the condition emulsified asphalt kept constant value 3.7%. The mixing water content was set to 0.8, 1.3, 1.8, 2.3, 2.8, 3.3, 3.8% respectively. Correspondingly, there were 7 emulsified asphalt cold recycled mixture because of 7 different mixing water content. The mold diameter was 150 mm and its gyratory force was 10.603 kN, so the work moving the per millimeter along axial direction is 10.603 J which is the effective work mean value. Test were results recorded in Table 2,

Table 2. Determination of the optimum total liquid usage based on the theory of effective work

Item	Total liquid dosage (%)						
	4.5	5.0	5.5	6.0	6.5	7.0	7.5
Displacement average (mm)	0.3097	0.3354	0.4258	0.4387	0.4677	0.4581	0.4354
The average of the effective work (J)	3.28	3.56	4.51	4.65	4.96	4.86	4.62

and the curve of total liquid content and effective power mean fitting relationship was shown in Fig. 2.

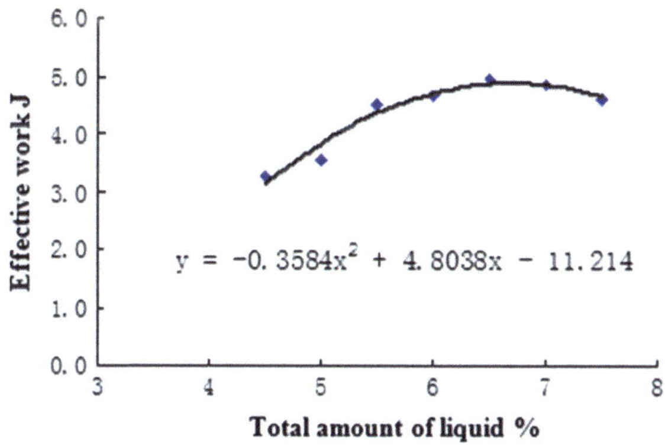

Fig. 2. The relationship between the total liquid mass and the average of the effective work

Test results showed that the effective work curve with the total amount of liquid has a peak. And based on the corresponding fitting curve and its equation, 6.5% was the best total amount of liquid.

3.5 Determination of Emulsified Asphalt Content

3.5.1 Initial Emulsified Bitumen Content

The total liquid amount (6.5%) was kept as a constant, the emulsified asphalt content value of 3.7% was used as the median. Then, 5 contents of emulsified asphalt were set at ±0.5 and ±1%, which included 2.7, 3.2, 3.7, 4.2 and 4.7% respectively. After mixing, the material was kept indoor at 20 °C temperature for 2.5 h, which was simulated the condition of construction delay the maximum theoretical density of the emulsified asphalt cold recycling mixture was calculated as Eq. (3).

$$\gamma_{\mathrm{MTD}} = \frac{P_{\mathrm{T}}}{\frac{P_1}{\gamma_1} + \frac{P_2}{\gamma_2} + \cdots + \frac{P_n}{\gamma_n} + \frac{\mathrm{EA\%}}{\gamma_{EA}} + \frac{W\%}{\gamma_W}} \tag{3}$$

where γ_{MTD} is emulsified asphalt cold recycled mixture maximum theoretical density, g/cm^3; P_T is emulsified asphalt cold recycled mixture quality, %; P_n is variety of mineral aggregate blending ratio, %; γ_n is variety of mineral aggregate density, g/cm^3; EA% is emulsified asphalt blending ratio, %; γ_{EA} is emulsified asphalt density, g/cm^3.

During the compaction of specimens, the height of them was recorded, and the bulk density G_{wb} was calculated. The range of targeted porosity is from 7 to 9%. The fitting curve was developed based on the relationship between the amount of emulsified asphalt and the porosity, on which the optimum content range of emulsified asphalt was determined.

The porosity is calculated as:

$$VV = (1 - \frac{G_{wb}}{\gamma_{MTD}}) \times 100\% \quad (4)$$

where G_{wb} is the compacted bulk density of EACRM, g/cm^3; γ_{MTD} is the theoretical maximum density of EACRM, g/cm^3.

Based on the test results shown in Table 3, the relationship between the amount of emulsified asphalt and porosity was drawn in Fig. 3.

Table 3. Target porosity determine emulsified asphalt usage test

Item	Emulsified asphalt dosage (%)				
	2.7	3.2	3.7	4.2	4.7
Density of the wet volume (g/cm^3)	2.227	2.265	2.273	2.306	2.303
Theoretical maximum wet density (g/cm^3)	2.514	2.495	2.476	2.483	2.464
Porosity (%)	11.4	9.2	8.2	7.1	6.5

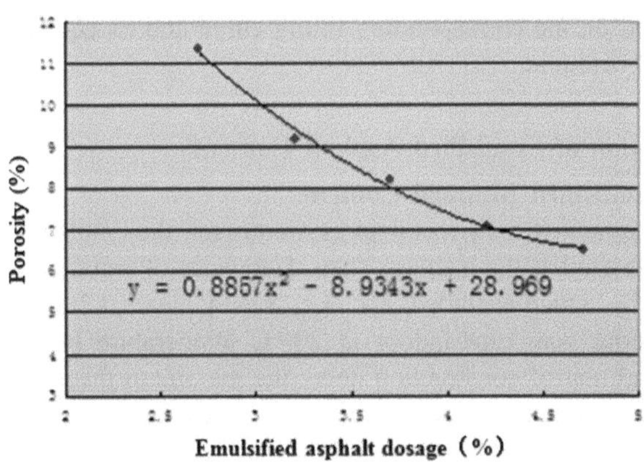

Fig. 3. The relationship between the amount of emulsified asphalt and the porosity

3.5.2 Mechanical Performances

After specimens were removed from the mold, they were cured in an air dry oven at 30 °C, and the change of weight was recorded. Curing of specimens was stopped until no change of specimens weight. The splitting strength test was used to evaluate the road performances of EACRM. The strength of the specimen shall be equal or greater than 0.5 MPa to meet specification requirement. The splitting strength test results were shown in Table 4. The relationship between the amount of emulsified asphalt and the splitting strength is shown in Fig. 4.

Table 4. Results of splitting strength test

Item	Emulsified asphalt dosage (%)				
	2.7	3.2	3.7	4.2	4.7
The average of the cleavage strength at room temperature 15 °C (MPa)	0.462	0.502	0.579	0.642	0.614

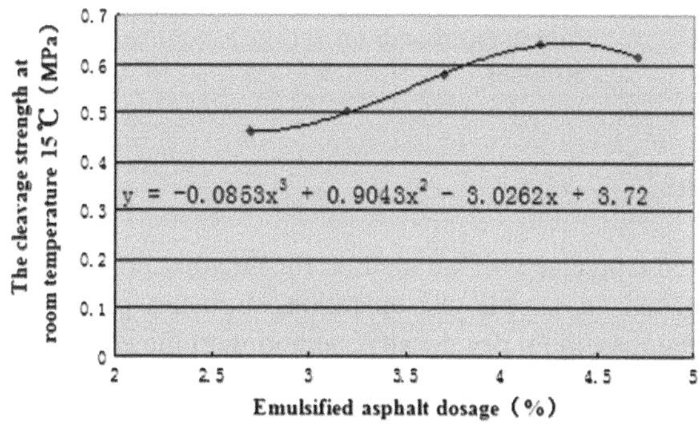

Fig. 4. The relationship between the amount of emulsified asphalt and the splitting strength

When other conditions are constant, the splitting strength of the mixture had a peak with the increasing of the amount of emulsified bitumen. The amount of emulsified asphalt was determined by the splitting strength index. When the amount of emulsified asphalt was between 3.2 and 4.2%, the splitting strength of all specimens were greater than 0.5 MPa, which met specification requirement. It confirmed that 7–9% of targeted porosity range was appropriate for emulsified asphalt recycling mixture design.

Based on the mixture porosity, mechanical performance and economics, it concluded that 3.7% is the optimum amount of emulsified asphalt.

4 Mixture Performance Test

Based on the mixture design (gradation design, 3.7% emulsified asphalt and 2.8% water), the recycling mixture specimens were mixed and compacted for the mixture performance test. Moisture susceptibility and high temperature stability test results were shown in Table 5. The results showed that these recycling mixture performances met the specification requirements. This indicated that the design method was feasible.

Table 5. Performance verification technical indicators

Test item		Test results	Technical requirements	Test method [8]
60 °C dynamic stability (time/mm)	Health 48 h	4310	≥ 3000	JTG E20 (T0719)
Freeze-thaw splitting strength ratio TSR (25 °C)	The specimen is double-faced with 50 times	83%	≥ 70% [7]	JTG E20 (T0729)

5 Conclusions

The paper proposed a precise volume method for the mix design of EACRM. The mix design was conducted via simple test operation, short test period and accurate water content. It used, the maximum dry density method with the effective compaction work to determine the optimum liquid dosage. The shear gyration compaction was used to simulate compaction method in the construction place, instead of heavy compaction test. The design method estimated the amount of emulsified asphalt by the regression equation, which is more representative than the Marshall design method that only has stable porosity of 4%. Based on the performance test of EACRM, it has better water stability and high temperature stability, which proved the method was a good design method for EACRM.

References

1. National Center for Asphalt Technology (NCAT): Participant's reference book. Pavement recycling guidelines for state and local governments [R]. U.S. Department of Transportation, Federal Highway Administration, Publication No. FHWA-SA-98-042 (1997)
2. Shi, F., Li, X., Sun, D., Lu, W.: Overview of the design method of cold recycled asphalt mixture. Highway **11**, 102–107 (2004)
3. Wu, C., Zeng, M., Zhong, M., Xiao, J.: Experimental study on design method of emulsified asphalt cold reclaimed mixture. J. Hunan Univ. **8**, 19–23 (2008)
4. Lu, W.: Regeneration Technology of Asphalt Pavement [M]. China Communications Press, Beijing (1989)
5. Wu, K., Li, Y., Yang, G.: Study on emulsified asphalt cold recycled asphalt mixture. J. Jinan Univ. (Nat. Sci. Ed.) **29**(3), 281–285 (2008)
6. Tan, Y., Dong, Z., Cao, L., Yuhui, E.: Application of Superpave volume design method to design cold recycled asphalt mixture. J. Highw. Transp. Res. Dev. **22**(3), 32–34 (2005)
7. JTG F41-2008: Technical specification for road asphalt pavement regeneration [S]
8. JTG E20-2011: Road engineering asphalt and asphalt mixture test procedures [S]

Evaluation of Rutting and Cracking Resistance of Foamed Warm Mix Asphalt Containing RAP

Mohammad Ashiqur Rahman[1(✉)], Amir Arshadi[2], Rouzbeh Ghabchi[3], Syed Ashik Ali[1], and Musharraf Zaman[1]

[1] University of Oklahoma, Norman, OK, USA
{marahman, syed.a.ali, zaman}@ou.edu
[2] AECOM, Chelmsford, MA, USA
amir.arshadi@aecom.com
[3] South Dakota State University, Brookings, SD, USA
Rouzbeh.Ghabchi@sdstate.edu

Abstract. Significant efforts have been made in recent years to construct sustainable and environment-friendly transportation infrastructure. These efforts are targeted toward saving natural resources, conserving the environment and reducing energy consumption. The use of Warm Mix Asphalt (WMA) technologies and increased utilization of Reclaimed Asphalt Pavement (RAP) in asphalt mixes are widely practiced by the asphalt industry. In the present study, the rutting and cracking resistance of WMA mixes containing RAP were evaluated in the laboratory and compared with those of Hot Mix Asphalt (HMA) containing the same amount of RAP. To assess the rutting resistance, Hamburg Wheel Tracking (HWT) tests were performed. Fatigue cracking performance was evaluated using Semi Circular Bending (SCB) test and Indirect Tensile Strength (ITS) test. The HWT tests data indicated that for similar numbers of wheel passes, higher rut depths were observed in the WMA specimens compared to the HMA specimens. This difference became increasingly dominant with a higher number of passes. Also, the WMA specimens exhibited more moisture-induced damage potential as indicated by the presence of stripping inflection point than the HMA specimens, as expected. From SCB and ITS tests, it was concluded that the fatigue and cracking resistances of foamed WMA specimens containing RAP were similar to those of their HMA counterparts. The findings of this study are expected to evaluate the laboratory performance of foamed WMA containing RAP.

Keywords: Eco-friendly flexible pavements · Infection point · Moisture-induced damage · Creep slope · Fatigue resistance

1 Introduction

Construction of sustainable and environmental-friendly transportation infrastructures results in saving natural resources, conserving the environment and reducing energy consumption. For more than a decade, as a part of efforts toward establishing a

D. Chen et al. (eds.), *Advancements on Sustainable Civil Infrastructures*,
Sustainable Civil Infrastructures, https://doi.org/10.1007/978-3-319-96241-2_11

sustainable and eco-friendly construction practice, the pavement industry has been using Reclaimed Asphalt Pavement (RAP) and various Warm Mix Asphalt (WMA) technologies in the production of asphalt mixes and construction of flexible pavements.

The WMA technologies are intended to reduce the viscosity of the asphalt binder using chemical additives, organic additives, and water-based or water-containing foaming processes (Alhasan et al. 2014). As a result, these technologies reduce the production temperature of traditional Hot Mix Asphalt (HMA) by about 20–40 °C (Rubio et al. 2012). Kristjansdottir (2014) suggested that about a 25–70% reduction in energy consumption can be attained compared to HMA due to low production temperature. There are several other benefits of using warm mix asphalt. Those include longer paving seasons, quicker turnover time to traffic, and improved working conditions due to lower odor, fume, and emission levels due to less heating of the asphalt binder. The WMA technologies also enhance compactability, reduce oxidative hardening of binders and reduce cracking in pavements (Rubio et al. 2012; Hurley and Prowell 2006; Gandhi et al. 2009).

Among all WMA technologies, the plant foaming technique (*called "foamed-WMA" in this paper*) has gained the most attention. Foamed-WMA uses water for foaming, which is more cost effective since there is no need to change the mixing process or to use any chemical additives. Moreover, with increasing asphalt binder cost and the scarcity of high quality aggregates, the demand for using RAP in asphalt mixes has increased rapidly in recent years. Due to economic and environmental benefits, a large percentage of WMA and HMA currently uses RAP. Although, using foamed WMA with RAP has many economic and environmental benefits, laboratory and field performance of such mixes is lacking. The present study addresses this weakness.

The volumetric properties of foamed WMA mixes can be similar to those of their HMA counterparts; however, the compactability, stripping resistance, rutting resistance and fatigue resistance of WMA mixes can be significantly different (Bonaquist 2011). With the addition of RAP in asphalt mixes, these characteristics become more difficult to predict. Zhao et al. (2013) found that the rutting resistance of WMA mixes was lower than the HMA mixes regardless of the WMA technology, RAP content, and structural layer. An increase in the rutting resistance was observed with the addition of RAP for both WMA and HMA specimens. Guo et al. (2014) observed that the fatigue life of the WMA mix decreased with the addition of RAP. The present study was undertaken to evaluate the rutting and cracking performance of WMA compared to HMA, where both containing a similar amount of RAP.

2 Objectives

The objectives of this study were to:

- Compare the rutting resistance of foamed WMA with that of hot mix asphalt (HMA) containing the same amount of RAP.
- Compare the cracking resistance of WMA with that of HMA containing the same amount of RAP.

3 Materials and Methodology

3.1 Materials

In this study, a commonly used paving mix (S4 mix with a nominal maximum aggregate size (NMAS) of 12.5 mm) for road construction in Oklahoma was selected to evaluate the rutting and cracking resistance of foamed WMA. It comprised of a PG 64-22 binder. The amount of water used for foaming the binder in the WMA production process was 2.0% by the weight of the binder. The properties of the aggregates are shown in Table 1. The mixes used in this study contained 5% fine RAP because of common practice for S4 mixes.

Table 1. Properties of aggregates

Aggregate	Producer/supplier	% Used
5/8″ chips	Hansen aggregates (Davis, OK)	35
Man. sand	Martin-Marietta (Davis, OK)	36
3/16″ chips	Hansen aggregates (Davis, OK)	10
Sand	General materials (Oklahoma City, OK)	14
Fine RAP	Contractor/project site	5

3.2 Mix Design

The mix design sheet was developed according to the AASHTO R35 design procedure (AASHTO 2013). The prepared mixes were compacted in a Superpave Gyratory Compactor (SGC) using 50 gyrations. The WMA mixes were produced in the laboratory using a procedure that simulates the conditions prevailing in a typical WMA plant in terms of mixing temperature and aging. Accordingly, the RAP was aged at a lower temperature of 135 °C. The duration of aging was kept similar to that of the HMA mix because similar aging duration is commonly used for both WMA and HMA in the asphalt production plant. The aggregates used in the WMA mix were also heated at a temperature of 135 °C, which is similar to the aging temperature of RAP. The selected binder was used to produce foamed-binder in the laboratory using an accu-foamer.

3.3 Hamburg Wheel Tracking (HWT) Test

The HWT tests were conducted on HMA and WMA specimens using the AASHTO T 324 test method (AASHTO 2016) and used to determine their rutting susceptibility and moisture-induced damage potential. In this study, the HWT tests were conducted at 50 °C with a wheel passing frequency of 52 passes/min and a wheel load of 705 N. The average linear speed of the wheel was approximately 1.1 km/h. The wheel traveled approximately 230 mm (9.05 in.) before reversing the direction. The test was automatically terminated after reaching a maximum rut depth of 20 mm or 20,000-wheel passes, whichever reached first. Deformations were measured along the length of the wheel path at 11 equally-spaced points. The rut depth at the mid-point of the sample was considered for further analysis. From the HWT test results, the post-compaction consolidation, creep slope, stripping slope, and stripping inflection point were determined.

3.4 Semi-circular Bend (SCB) Test

Cracking is one of the major distresses in flexible pavements. It is primarily caused by repeated traffic loading, and can lead to substantial reduction in serviceability of flexible pavements (Shu et al. 2008). To determine the cracking resistance of asphalt mixes, Louisiana SCB test procedure was followed in this study. This method characterizes the cracking resistance of asphalt mixes at an intermediate temperature (20 °C in this study) and in terms of the critical strain energy release rate (J_c). As shown in Eq. (1), J_c is the strain energy release rate per notch depth. A higher J_c value represents a higher resistance to cracking.

$$J_c = -\left(\frac{1}{b}\right)\frac{\mathrm{d}U}{\mathrm{d}a} \tag{1}$$

where
- b SCB specimen thickness;
- a notch depth; and
- U strain energy (area under stress–strain curve).

The SCB tests were conducted on half-disk-shaped specimens having a diameter of 150 mm (6 in.) and a thickness of 50 mm (2 in.). In order to determine the critical value of J-integral (J_c) using Eq. (1), the SCB tests are generally conducted on specimens with at least two different notch depths. In this study, three different notch depths of 25.4, 31.8, and 38 mm were selected to increase the measurement accuracy by developing a linear regression correlation for the strain energies versus notch depths for each mix. The specimens were loaded monotonically at a rate of 0.5 mm/min using a three-point flexural apparatus (Kim et al. 2012), as shown in Fig. 1.

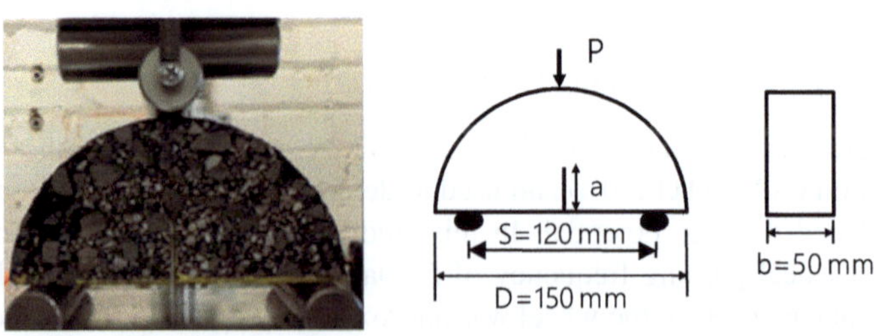

Fig. 1. Test setup for semi-circular bending (SCB) testing (Kim et al. 2012)

3.5 Indirect Tensile Strength (ITS) Test

To determine fatigue cracking resistance of asphalt mixes, the Indirect Tensile Strength (ITS) test was used. This test was conducted according to ASTM D 6931 test method (ASTM 2010). The ITS specimens were prepared in the laboratory using a Superpave® gyratory compactor. The diameter and height of the compacted specimens were 150 mm (6 in.) and 120 mm (4.8 in.), respectively. The compacted specimens with an air void of 7 ± 0.5% were selected for conducting the ITS tests. The indirect tensile strength was calculated using the following equation:

$$S_t = \frac{2000 * P}{\Pi * t * D} \quad (2)$$

where

S_t = indirect tensile strength; P = maximum load; t = specimen height immediately before test; and D = specimen diameter (ASTM 2010).

4 Results and Discussion

4.1 Rutting Resistance

As a supplement to mix design procedure, many transportation agencies have started using Loaded Wheel Testers (LWT) to evaluate the rutting potential of asphalt mixes. The results obtained from these wheel tracking devices have been found to correlate well with the actual field performance when the loading and environmental conditions of a given location were considered (Miller et al. 1995; Cooley et al. 2000; Kandhal and Cooley 2002). Among the LWTs available, the Hamburg Wheel Tracking (HWT) test is gaining popularity due to its ability to screen mixes for rutting and simulating field performance data reasonably well.

Figure 2 shows the HWT results obtained in this study for both foamed WMA and HMA mixes containing an identical amount of RAP. In Table 2, rut depths at 1000, 5000, 10,000, 15,000, and 20,000 wheel passes for all mixes are presented. From Fig. 2 and Table 2, it is seen that higher rut depths were observed for WMA specimens compared to the HMA specimens for similar wheel passes. These differences became more significant for higher numbers of wheel passes. For 20,000 wheel passes, a rut depth of more than 10 mm was observed for the foamed WMA specimens containing RAP. Moreover, from Fig. 2 it is noted that for WMA mixes, there was a distinct inflection point at about 12,000 wheel passes, which indicated potential for moisture-induced damage in the WMA specimens. Consequently, after 12,000 wheel passes, the rate of rut depth increased rapidly. Moreover, the WMA specimens were more prone to moisture-induced damage in the field. Comparatively, HMA specimens did not exhibit any stripping inflection points.

The performance parameters determined from the HWT curves are presented in Table 3. The creep slope for the WMA mixes is found to be 2946 passes/mm, whereas, for the HMA mixes this creep slope is about 15,000 passes/mm. Thus, the HMA specimens showed about five times higher resistance to creep than the WMA

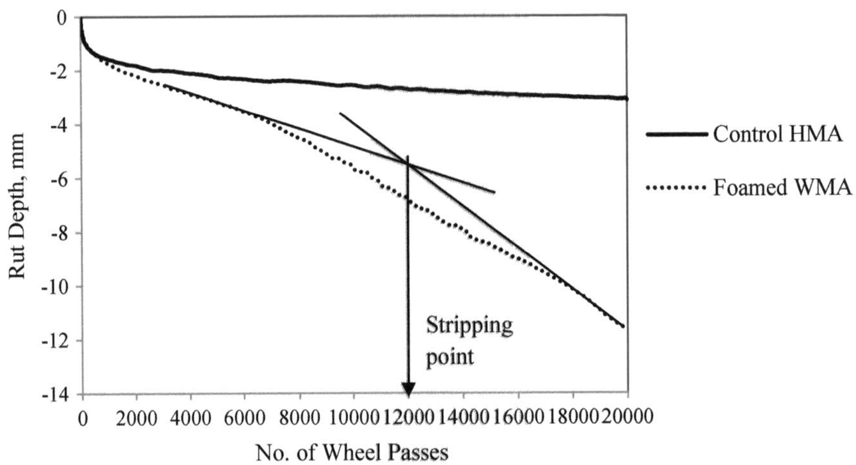

Fig. 2. Comparison of HWT test results for WMA and HMA mixes

Table 2. Comparison of rut depths (mm) for foamed WMA and HMA mixes at different numbers of wheel passes

MIX type	Wheel passes				
	1000	5000	10,000	15,000	20,000
HMA S4	1.58	2.25	2.55	2.88	3.11
WMA S4	1.76	3.17	5.6	8.51	11.56

specimens. These results also indicated that the foamed WMA specimens became more susceptible to rutting and moisture-induced damage compared to their HMA counterparts containing an identical amount of RAP. Hence, despite of great economic and environmental benefits, rutting resistance remains a matter of concern for foamed WMA containing RAP. This is partly attributed to a smaller degree of binder aging during the production of mixes because of lower production temperature for foamed WMA. Similar observations have been reported by Wielinski et al. (2009), Hill (2011), Ali et al. (2013), and Zhao et al. (2013).

4.2 Cracking Resistance

In the present study, the SCB and ITS tests were conducted to evaluate the cracking resistance of selected mixes. From Fig. 3, it is observed that both WMA and HMA SCB specimens showed a similar average strain energy for a specific notch depth. The strain energy is found to decrease with an increase in notch depth. The difference between average strain energy for a specific notch depth was not significant at 95% confidence level. Table 4 represents a summary of two-tail t-test at 95% confidence level for the difference in the strain energy between the HMA and WMA specimens. For all notch depths, the p-value was much higher than 0.05. So, the

Table 3. Performance parameters obtained from HWT test results for foamed WMA and HMA mixes

MIX type	HWT indices				
	Post compaction (mm)	Creep slope (passes/mm)	Stripping inflection point (passes)	Inverse stripping slope (mm/pass)	Stripping slope (passes/mm)
HMA S4	1.58	15,000	N/A	N/A	N/A
WMA S4	1.76	2946	12,000	0.0007556	1323.53

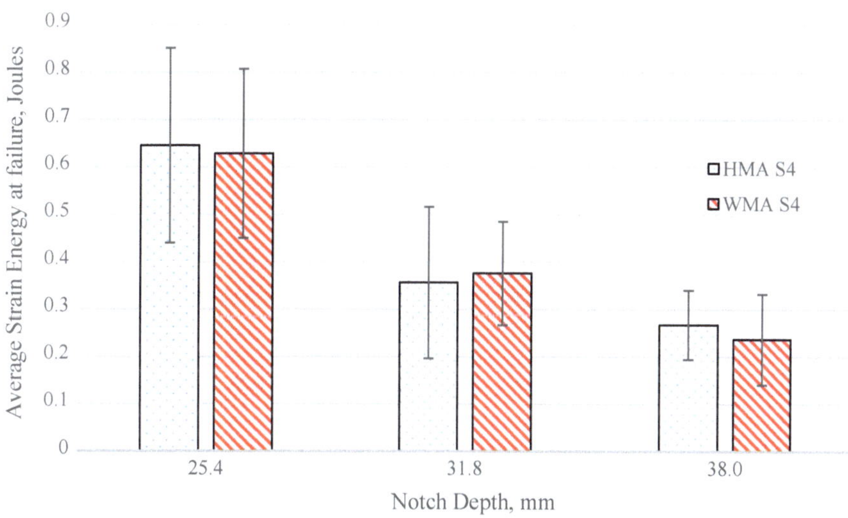

Fig. 3. Variations in average strain energy at failure with notch depth, Joules versus notch depth, mm

difference in average strain energy between the HMA and WMA specimens was not significant at 95% confidence level for a specific notch depth.

The J_c values obtained from the average strain energy for foamed WMA and HMA specimens containing RAP are presented in Fig. 4. The critical strain energy release rate for each mix was calculated using Eq. (1). The average J_c value for the foamed WMA specimens was 0.630 kJ/m^2, whereas the average J_c value for the control HMA was 0.597 kJ/m^2. Thus, foamed WMA showed a slightly higher resistance to cracking than the HMA containing the same amount of RAP. Dong et al. (2017) also observed that foamed WMA had more fatigue resistance than HMA with the same amount of RAP.

From the ITS test results, a similar cracking resistance was observed for both foamed WMA and HMA specimens containing the same amount of RAP. The average

Table 4. Summary of t-test for the difference in the average strain energy

Notch depth (mm)	25.4		31.8		38	
Mix type	HMA S4	WMA S4	HMA S4	WMA S4	HMA S4	WMA S4
Mean	0.64	0.63	0.36	0.38	0.27	0.24
Variance	0.04	0.03	0.03	0.01	0.01	0.01
t Stat	0.13		−0.21		0.57	
P(T ≤ t) two-tail	**0.90**		**0.84**		**0.59**	
t Critical two-tail	2.45		2.57		2.31	

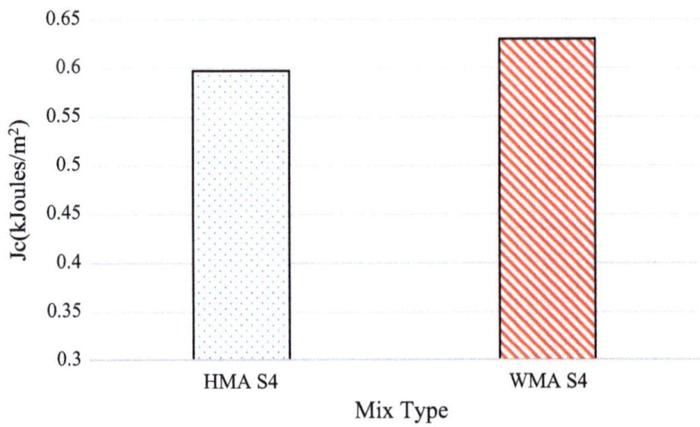

Fig. 4. Comparison of *J*-integral (*Jc*) values for different mixes

Indirect Tensile Strength (ITS) was found to be 3760 kPa for the foamed WMA and 3740 kPa for the HMA. So, the difference between the average indirect tensile strength of WMA and HMA is statistically insignificant at 95% confidence level, as shown in the Table 5. Thus, based on the SCB and ITS test results, both foamed WMA and HMA showed a similar level of resistance to cracking (Fig. 5).

Table 5. Summary of t-test for the difference in the indirect tensile strength t-test

	HMA S4	WMA S4
Mean	3440	3460
t Stat	−0.05	
P(T ≤ t) two-tail	**0.97**	
t Critical two-tail	4.30	

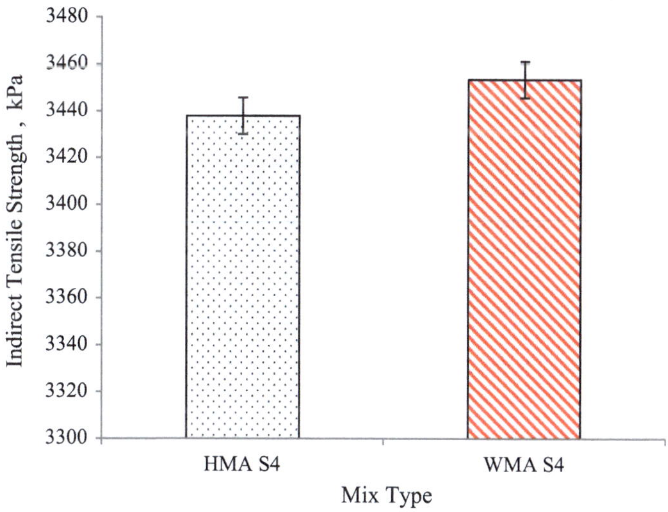

Fig. 5. Indirect tensile strength (ITS) test results

5 Conclusions

In this study, laboratory performance of foamed WMA and HMA containing an identical amount of RAP was evaluated. From HWT tests, higher rut depths were observed in foamed WMA mixes than HMA mixes for similar number of wheel passes. These differences increased with higher number of passes. After 12,000 wheel passes striping inflection points were observed in case of foamed WMA specimens indicating possibility of moisture-induced damage for these specimens. Moisture-induced damage was not observed for HMA. In terms of cracking both foamed WMA and HMA mixes showed similar resistance, with statistically insignificant differences.

Acknowledgements. The authors are thankful to the Southern Plains Transportation Center (SPTC) for providing financial support of this study. Financial support of the Oklahoma Department of Transportation (ODOT) is also gratefully acknowledged. The authors are also thankful to Silver Star Construction Co. Inc. for providing the necessary materials for this study. Finally, the authors acknowledge the support of co-workers in Broce Laboratory for their assistance throughout this study.

References

Alhasan, A.A., Abbas, A.R., Nazzal, M., Dessouky, S., Ali, A., Kim, S.S., Powers, D.: Low-temperature characterization of foamed warm-mix asphalt produced by water injection. Trans. Res. Rec.: J. Trans. Res. Board (2445), 1–11 (2014). Transportation Research Board of the National Academies, Washington, DC

Ali, A., Abbas, A., Nazzal, M., Alhasan, A., Roy, A., Powers, D.: Effect of temperature reduction, foaming water content, and aggregate moisture content on performance of foamed warm mix asphalt. Constr. Build. Mater. **48**, 1058–1066 (2013)

American Association of State Highway and Transportation Officials: AASHTO R-35 standard practice for Superpave volumetric design for asphalt mixtures. Transportation Research Board, National Research Council, AASHTO, Washington, DC (2013)

American Association of State Highway and Transportation Officials: AASHTO T-324 standard method of test for Hamburg wheel-track testing of compacted hot mix asphalt (HMA). Transportation Research Board, National Research Council, AASHTO, Washington, DC (2016)

American Society for Testing Materials International: ASTM D6931 standard test method for indirect tensile (IDT) strength of bituminous mixtures. ASTM, West Conshohocken, Pennsylvania (2010)

Bonaquist, R.F.: NCHRP (National Cooperative Highway Research Program) Report 691, Mix Design Practices for Warm Mix Asphalt. Transportation Research Board, Washington, DC (2011)

Cooley, L.A., Kandhal, P.S., Buchanan, M.S., Fee, F., Epps, A.: Loaded wheel testers in the United States: state of the practice. In: Transportation Research Circular, E-C016. Transportation Research Board, Washington, DC (2000)

Dong, F., Yu, X., Xu, B., Wang, T.: Comparison of high temperature performance and microstructure for foamed WMA and HMA with RAP binder. Constr. Build. Mater. **134**, 594–601 (2017)

Gandhi, T., Akisetty, C., Amirkhanian, S.: Laboratory evaluation of warm asphalt binder aging characteristics. Int. J. Pavement Eng. **10**, 353–359 (2009)

Guo, N., You, Z., Zhao, Y., Tan, Y., Diab, A.: Laboratory performance of warm mix asphalt containing recycled asphalt mixtures. Constr. Build. Mater. **64**, 141–149 (2014)

Hill, B.: Performance evaluation of warm mix asphalt mixtures incorporating reclaimed asphalt pavement. Ph.D. thesis, University of Illinois at Urbana-Champaign, Champaign, IL (2011)

Hurley, G., Prowell, B.: Evaluation of potential process for use in warm mix asphalt. J. Assoc. Asphalt Paving Technol. **75**, 41–90 (2006)

Kandhal, P.S., Cooley, L.A.: Evaluation of permanent deformation of asphalt mixtures using loaded wheel tester. Asphalt Paving Technol. **71**, 739–753 (2002)

Kim, M., Mohammad, L., Elseifi, M.: Characterization of fracture properties of asphalt mixtures as measured by semicircular bend test and indirect tension test. Trans. Res. Rec.: J. Trans. Res. Board **2296**, 115–124 (2012). https://doi.org/10.3141/2296-12

Kristjansdottir, O.: warm mix asphalt for cold weather paving. Ph.D. thesis, University of Washington, Seattle, WA (2014)

Miller, T., Ksaibati, K., Farrar, M.: Using Georgia loaded-wheel tester to predict rutting. Trans. Research Rec.: J. Trans. Res. Board **1473**, 17–24 (1995)

Rubio, M.C., Martínez, G., Baena, L., Moreno, F.: Warm mix asphalt: an overview. J. Clean. Prod. **24**, 76–84 (2012)

Shu, X., Huang, B., Xiang, S., Vukosavljevic, D.: Laboratory evaluation of fatigue characteristics of recycled asphalt mixture. Constr. Build. Mater. **22**, 1323–1330 (2008)

Wielinski, J., Hand, A., Rausch, D.M.: Laboratory and field evaluations of foamed warm-mix asphalt projects. Trans. Res. Rec.: J. Trans. Res. Board (2126), 125–131 (2009). Transportation Research Board of the National Academies, Washington, DC

Zhao, S., Huang, B., Xiang, S., Woods, M.: Comparative evaluation of warm mix asphalt containing high percentages of reclaimed asphalt pavement. Constr. Build. Mater. **44**, 92–100 (2013)

Review and Assessment of Flexible Pavement

Sabir S. Sayyed[1], R. P. Patil[2], Anand Tapase[3(✉)], A. C. Attar[4], and P. G. Chandak[1]

[1] Visvesvaraya Technological University, Belagavi, India
sssabirsayyad@gmail.com, chandak.p.88@gmail.com
[2] Jain AGMIT, Jamkhandi, India
ravindrappatil@gmail.com
[3] Rayat Shikshan Sanstha's, Karmaveer Bhaurao Patil College of Engineering, Satara, India
tapaseanand@gmail.com
[4] Rajarambapu Institute of Technology, Sakhrale, India
abdulrashid.attar@ritindia.edu

Abstract. The failure of flexible pavement is termed in a cumulative number of standard axles required to cause a pavement to rut up to a depth of 20 mm or the number of standard axles required to exceed fatigue cracking beyond 20% of the pavement surface area as per the Indian code of practice. This study illustrates the effectiveness of finite element analysis in considering a practical design approach for achieving the optimum pavement section subjected to loading. From the obtained results, it is noticed that the value of horizontal tensile strain increases at each trial increase in temperature, showing the adverse effect of a rise in temperature on the life of the pavement. When the stiffer binder is used considering the decrement in temperature of the underneath layer, a drastic increase in the value of tensile strain is noticed which is at much higher side than the allowable limits. The observed value indicates that the selected combination will give rise to fatigue cracking. When the increment in temperature is considered then the reduction in the value of tensile strain even up to 35% is noticed. From the analysis, it is concluded that the use of too soft bituminous mix results in lowering the structural ability of the pavement at high temperatures and too hard bituminous mixes would become brittle at low temperatures resulting in cracking under loading.

Keywords: FEM · Flexible pavement

1 Introduction

The current design procedures for flexible pavements are based on past knowledge and in situ tests data like American Association of State Highway Official's (AASHO) Road Test. Chandak et al. (2018) observed that the basic categories on which the current flexible pavement design methods are dependent are limiting shear failure methods, empirical methods, regression methods, limiting deflection methods. The inconvenience of an empirical approach like CBR method of design is that the result is dependent on material, environmental and loading conditions. Limiting shear failure

D. Chen et al. (eds.), *Advancements on Sustainable Civil Infrastructures*,
Sustainable Civil Infrastructures, https://doi.org/10.1007/978-3-319-96241-2_12

approach fails to deal with riding convenience in addition to safety and stability parameters. While the limiting deflection approach fails to address the pavement distresses like potholes, fatigue, and rutting adequately. The design equations based on AASHTO method can be helpful to the situations for which it was designed. Various factors affect design and performance of flexible pavement, while current practices for design of thicknesses of different layers of flexible pavement mainly consider CBR of subgrade and traffic i.e. (cumulative standard axle load) neglecting a number of parameters like temperature (AASHTO 1993). Chakraborty and Das (2003) states that the flexible pavement usually consists of four to five layers of different materials, thicknesses and material characteristics making it more complex for its analysis.

To meet the growth of industrial development road density has been increased in many of the developing country like India. These increased infrastructure facilities have to cater traffic on these roads for their changing static and dynamic characteristics of vehicle, overloaded vehicles. Ranadive and Katkar (2010), Ranadive and Tapase (2013) and Tapase et al. (2017) identified these factors which are responsible for early deterioration of pavement in terms of pot holing, cracking, undulation, raveling, rutting etc.in bituminous layer. Instrumented pavement is one of the technique that can be used to overcome this problem by knowing the effect of these overloaded vehicles, but as these locations are widely spaced throughout the country making this process tedious and time consuming. So this issue can be addressed by preparing proper design charts supported with proper design procedure to fulfill the present day requirements.

2 Literature Review

Mechanistic pavement response models developed over the years are Boussinesq's one-layer model, multi-layer elastic theories, and finite element models. Boussinesq developed a formulation to calculate stresses, strains, and deflections of homogeneous, isotropic, linear elastic, with Young's modulus and Poisson's ratio exposed to a static point load. Boussinesq's equations were later expanded by researchers for a uniformly distributed load by integration. Yoder and Witczak (1975) recommended Boussinesq theory to be used to evaluate subgrade stresses, strains, and deflections when the base and the subgrade have same stiffness. Burmister developed a solution for two and three-layered system. Flexible pavement responses can be modeled using finite element analysis. Yang (2008) categorized design factors which should be integrated into traffic and loading factors. Elastic moduli of different layers are affected by temperature and seepage which in turn affect the pavement performance. Environmental factors like the nearness of ground water table, frost infiltration, freezing index are the factors needed to be considered in the analysis. In mechanistic design procedure, the heat transfer model for evaluating the temperature distribution; the moisture equilibrium model for evaluating moisture distribution in the subgrade, and penetration-drainage model can be implemented. Helwany et al. (1998) studied 3 layered pavement system and applied various loading conditions which comprised of various tyre pressure, different axle load configuration with different types of materials like elastic, non-elastic etc. The study proposed a need for a model which will be helpful to predict pavement performance by finite element method after its validation.

Ghosha et al. (2013) studied the current practice of flexible pavement design based on IRC-37, considering the effect of variation in speed of vehicles, pavement temperature and aging of material on performance which were previously ignored even these factors have impacts on the bituminous mixture and henceforward the pavement thickness. In this investigation, the effect of the various parameter along with dynamic modulus of the bituminous mixture was considered and evaluated using AASHTO-WARE software. It was found that pavement cross section failed to fulfill the design necessities when exposed to rigorous examination. For the expected design life, the thickness of the bituminous layer was found which was found very high when compared with existing pavement cross sections.

Tapase et al. (2016, 2017) showcased usefulness of finite element analysis to evaluate critical parameters for varying thicknesses of pavement which shows the decrement in tensile strength and compressive strain at critical locations for increasing bituminous layer thicknesses. Ranadive and Tapase (2016a, b) analyzed flexible pavement by using axisymmetric analysis considering various parameter for different thicknesses. It is observed deflection of pavement varies with variation in thicknesses of pavement and shows more deflection for varying base and sub base thicknesses.

Mulungye et al. (2007) highlighted the use of finite element method to calculate pavement distresses using ANSYS and modeling the pavement as 2 dimensional and four-layered structure. The flexible pavement was evaluated for transportation truck considering axle load variations, wheel configurations and tyre pressure on soft soil subgrade using finite element method and verifying the results with in situ field data and established the use of FEM to predict pavement responses to cyclic wheel loads accurately. Rahman et al. (2011) studied pavement response exposed to various traffic factors like different, tyre imprint, and axle configuration and inflation pressure. Finite element analysis was done about permanent deformation and fatigue using ABAQUS software. Three elastic layers of pavement were considered as bituminous surface, granular base and subgrade which were expected to reply linearly to applied static load. Interface properties were presumed such as frictionless contact between two adjacent layers. The analysis was done by ABAQUS software to find vertical compressive and horizontal tensile strains. The study shows that tyre imprint area should be a rectangle with semicircles on either side and more extensive research work has to be done in this area.

Su et al. (2008) stated that the current design methodology considering contact stress equal to tyre inflation pressure and being distributed uniformly over a circular area prove to be inaccurate. For better accuracy, the authors calculated the shear stress in asphalt mix layer by applying non uniform stresses which depicted the occurrence of maximum shear stress at approximately 60 mm under the edge of the tyre. Wang and Roque (2011) studied the effect of tyre pressure for various configurations near the surface by using finite element. Based on the results, pavement interaction models were developed which were further validated by comparison with measured values. The study showed that super single tyre produced more damage near the surface than upcoming wide-based tyres. Wang and Chong (2015) found that long life flexible pavement is a good solution for heavily trafficked flexible pavement. As after long service insufficient bonding and air voids together at interface layers are the major cause of deterioration and can be attended by considering some maintenance practices during design and construction of flexible pavement as per particular climatic condition.

Masad et al. (2006) studied the characterization of pavement layers as isotropic and anisotropic to predict the permanent deformation and fatigue of flexible pavements using NCHRP 1-37 A models. The study showed that high tensile stresses were observed in lower sides of asphalt layers causing permanent deformation in asphalt layers when it was considered in terms of anisotropic properties in NCHRP 1-37A model. Chiasson et al. (2008) used 2-dimensional finite element modeling linear approach to measure the distribution of temperature and thermal stresses related to it. Immanuel et al. (2008) used elastic analysis to predict stresses in base and subgrade layers of pavement, which were later compared with actual field values which showed some variation at higher pressures. The author reported that elastic analysis method is suitable for dynamic loading when it is considered with different environmental and loading conditions. The authors state that the study can further be extended to know the nonlinear behavior of materials, vertical pressure, and accuracy of parameters like asphalt strain.

Siddharthan et al. (2005) studied pavement behavior for varying environmental and loading conditions to find stresses by overloaded vehicles and developed design methodology to solve such complex loading condition due to bulky off-road vehicles. Animesh and Pandey (1999) studied the pavement design method in India, which does not consider the heavy traffic and different climatic conditions. Rutting and fatigue criteria were used to propose new design method with the mechanistic approach.

Yang et al. (2015) analyzed pavement performance by AASHTOWARE software for varying climatic inputs. The findings showed that performance of pavement is utmost sensitive to variations in temperature with other climatic factors. Diefenderfer et al. (2006) established a model for pavement temperature which is helpful to find temperature profile to determine pavement characteristics. The previous trends from records can predict future pavement temperature and the extent to which it is exposed to critical temperature. Zuo et al. (2007) studied the effect of varying temperature and water content on pavement life. The study showed that seasonal temperature variation, temperature gradient and moisture percentage in base and sub grade affected the assessment of pavement life.

Maadani et al. (2017) evaluated the capacity of the design steps to describe the impact of moisture content and temperature on the flexible pavement. The resilient modulus projected to describe unbound materials was more sensitive to moisture. The thickness limits involved in the Mechanistic Empirical Pavement Design Guide (MEPDG) does not affect performance when a thick layer is substituted with two layers of the same material. The developed model was sensitive to climatic variations. Rutting predictions for the warm region were about 50% greater than those for the colder regions, while the model had restricted sensitivity to unbound materials. Permanent deformation of base and subgrade materials presented nearly no difference amongst different states viz. dry, optimum, and wet. For further research, it is recommended to use nonlinear analytical methods to increase model's sensitivity as the current model works on linear elastic analysis with permanent deformation prediction. Hasan et al. (2015) developed the relationship between empirical and mechanistic empirical methods by considering various factors like temperature, precipitation and their effect in terms of different distress to form various forms of cracking. The study shows that longitudinal cracking is more sensitive to temperature and precipitation while the negligible impact on the expected International Roughness Index (IRI).

Sahaa et al. (2014) investigated the performance of flexible pavement for varying climatic conditions using MEPDG. It was observed that total pavement rutting, international roughness index (IRI) and asphalt concrete are amongst most sensitive parameters to climatic changes. Orobioa and Zaniewskib (2013) used a software tool for mechanistic-empirical pavement design as the bulky calculations required for the design of pavement. A huge number of inputs like material properties, traffic characteristics, and climatic condition are essential for MEPDG. The findings of the study are useful to decide the hierarchical order for traffic inputs depending on their sensitivity.

Gogoi et al. (2013) analyzed long-term pavement performance sections showing fatigue and rutting distress to study the correlation of both with each other. The research study established that rutting depth increased with fatigue cracking till a level of about 5% and afterward rutting stabilizes and does not show increment with an increase in cracking due to fatigue. El-Shaib et al. (2017) compared different approaches established on AASHTO 1993 and MEPDG. The difference between approaches was proved for predicting the performance of pavement when traffic increases and subgrade strength decreases for varying climatic conditions. For strong subgrade and low traffic climatic condition had a significant effect on distress particularly on AC rutting.

Saha et al. (2014) focused on MEPDG application in Alberta. The study highlighted future efforts essential for the betterment of the MEPDG flexible pavement performance models. A systematic and critical evaluation was made in between the pavement design accomplished succeeding the AASHTO-based Alberta Transportation Pavement Design (ATPD) method and using MEPDG. Six different design situations for different traffic levels and subgrade materials were comprised in the study. The evaluation was made through the investigation of MEPDG-predicted distress dependability at end of the design period for the ATPD pavement design thicknesses. For this analysis default values in MEPDG were taken for the reliability and distress criteria. The analysis discovered that, although a higher level of reliability is recommended in the ATPD method for stretches with high traffic intensity, pavement design thickness does not encounter MEPDG default necessities. Moreover, subsequent the ATPD method to design cases of poor subgrade materials effects in higher MEPDG distresses in evaluation to cases of strong subgrade materials. The determinations offered in the study want to be continuous towards more evaluation of the usefulness of both pavement design guides for the region. In this analysis, the performance of the studied pavement sections across the region, previously designed using ATPD, want to be fully assessed in relationship to design guidelines from the MEPDG.

Considering the available literature and the design guidelines for flexible pavements, it is observed that considerable progress has been made in the design procedure bearing in mind many factors which were previously ignored. It is seen that substantial progress has been made in relation to the use of finite element models, multilayered elastic theories etc. However, there are some factors which still need to be addressed in the designing process.

The design charts based on different values of CBR should be tested for thickness variation of different layers of the pavement section by considering variation in temperature of the underneath layers. As such keeping in mind both the important parameters like initial costs, maintenance costs, design life a suitable design can be suggested for various environmental, loading and local conditions. The design procedure is extensively drawn based on the findings of AASHTO for the USA climatic zones with consideration of reliability level but it is imperative to consider the different climatic conditions that affect the pavement for various regions. Further, the study on climatic conditions can well be extended to consider the effect of daytime temperature and nighttime temperature on the pavement and also due to seasonal changes along with decremental and incremental variation within the different component layers.

3 Pavement Composition and Methodology Adopted

To evaluate the average annual pavement temperature (AAPT) and number of correlations have been developed relating to average monthly pavement temperature (AMPT), average monthly air temperature (AMAT) along with average annual pavement temperature and average annual air temperature which may not be strictly valid for Indian conditions. Number of researchers have already reported low pavement temperature, unpredictable daily and cyclic temperature along with environmental factors are responsible for reducing life of the pavement hence in the present study a bituminous layer comprising of bituminous concrete (BC) and dense bituminous macadam (DBM) are considered separately whereas BC is the wearing course laid above the DBM. As per the guideline, the average annual pavement temperature is taken as 35 °C and the modulus of bituminous layer are defined for the upper layers by adopting the same temperature.

In India, higher summer temperature may increase beyond 45 °C and in some regions like Jammu & Kashmir, Himalayan region low temperature even below 0 °C is observed. So IRC 37:2012 (2012) is guideline prepared considering the average annual temperature to be around 35 °C which may not be validated for various regions within the country. The geographical location of country, the nearness of Indian Ocean to the south and Himalayan ranges to the north makes the job of a pavement engineer difficult in analyzing and designing a pavement for particular location. Hence it is proposed to have a separate design guideline in various parts, that is, IRC 37 in sub-parts for different regions of the country.

To design an economical and stable section considering the availability of conventional material, CBR value of sub-grade and the local construction practices should be taken into consideration. The thickness (h_1) of the bituminous layer is taken as 200 mm, wherein bituminous concrete (BC) top layer of bituminous mix is kept constant at 50 mm and DBM is taken as 150 mm. In all two trial conditions are analysed considering the atmospheric pressure to be constant at approximately 40 and 20 °C. When temperature of BC layer is assumed at 40 °C five cases of decrement from 40 °C up to 20 °C are analyzed by a decrement of 5 °C for each trial for Viscosity Grade (VG30) of bituminous mix as per IRC: 37-2012. Similarly, when the BC layer temperature is assumed to be 20 °C six cases of increment from 20 °C up to 40 °C are analyzed by a increment of 5 °C for each trial. The granular base (225 mm) and granular sub-base (225 mm) are treated as a single granular layer of 450 mm with crushed rock having material property as, E_3 = 450 MPa, μ_2 = 0.35 as a base layer material is kept constant throughout the analysis. A uniform pressure of 0.800 MPa (800 kPa) caused by the modern trucks is applied on a circular contact area having a radius of 150 mm as shown in Fig. 1. Modulus of various bituminous mixes recommended by IRC: 37-2012 guidelines as mentioned in Table 1 based on the results of extensive laboratory testing as per American Society of Testing Material. (ASTM D7369-09) throughout the country which are obtained at different temperatures from 20 to 40 °C (IRC 37-2012) are extracted for the study. All the above stated trials, are typically checked for their suitability in the pavement section for selected subgrade condition for E_3 = 80 MPa. If such type of analysis is validated, it will prove to be beneficial to derive useful design charts for any combinations of thicknesses, material properties and field conditions without relying on theoretical/empirical design procedures. These hypothetical thicknesses and material properties are generally considered in practice as per IRC: 37-2012; hence it is an attempt to correlate the present study with actual field conditions.

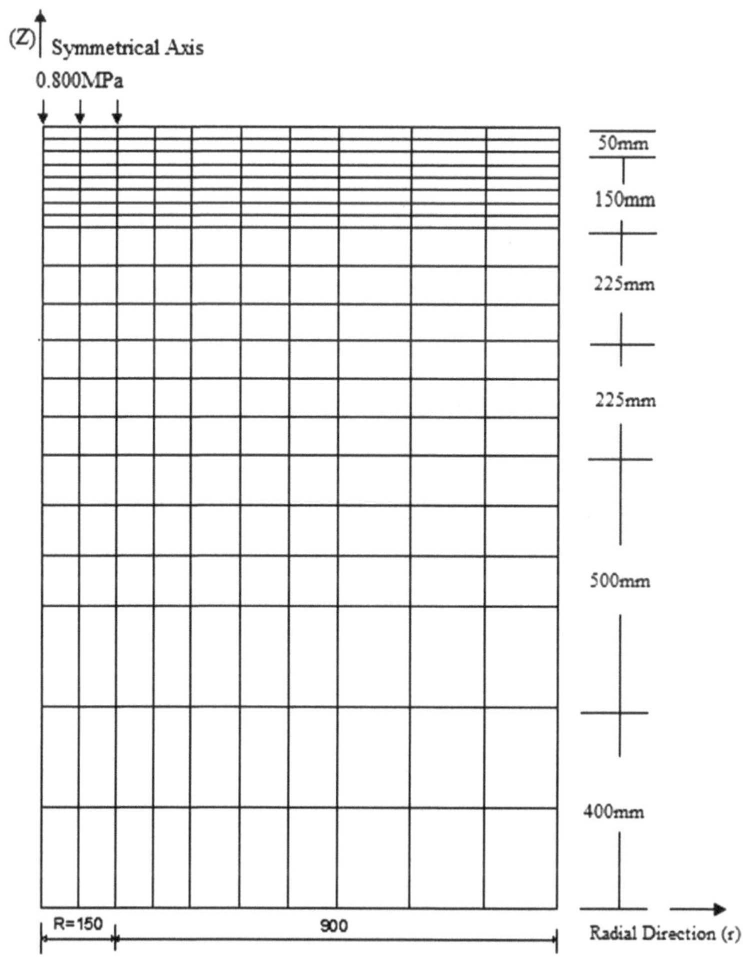

Fig. 1. Finite element idealization of the pavement section (All dimensions are in mm)

Table 1. Cases for analysis with mix type for viscosity grade (VG-30)

BC temperature	Modulus of DBM for VG30 at different temperatures (E_2)				
Decrement in temperature					
40 °C	40 °C	35 °C	30 °C	25 °C	20 °C
E_1 = 1250 MPa	1250 MPa	1700 MPa	2500 MPa	3000 MPa	3500 MPa
Increment in temperature					
20 °C	20 °C	25 °C	30 °C	35 °C	40 °C
E_1 = 3500 MPa	3500 MPa	3000 MPa	2500 MPa	1700 MPa	1250 MPa

Finite element technique follows three basic steps for the analysis, the first stage starts with investigation of system idealization followed by formulation of governing equation and their solution and last step is of evaluation of structural response for developing design process Tapase and Ranadive (2016). Pavement system is analysed by consideration of modulus of elasticity and poisons ratio of material being used for pavement construction. Various materials can be used for different layers of pavement and further combinations can be obtained to have good blend by extracting useful data from available literature. Zienkiewicz and Taylor (1991), Zeeveart (1982) idealization of finite element analysis is done by four nodes quadrilateral element. Discretization of pavement system into three layers with boundary fixation at 8 times the loaded radius. Sam et al. (1999), Sinha et al. (2014) fixed the right boundary more than 7 times the radius 150 mm of the applied load. The present work considers the total pavement thickness in between 1450 and 1600 mm, variation in bituminous layer in the range of 100–250 mm for 50 mm increment with sub layer thickness of 25 mm, base with constant thickness of 450 and 75 mm sublayer thickness, while subgrade is considered up to 900 mm thickness. Bottom of subgrade is restrained in both axial (z) and radial (r) directions as shown in fig. It is assumed extreme right boundary that is at a distance of 1050 mm from line of action of wheel load not undergo radial displacements, restraining nodes in radial (r) directions. In continuation with the investigation and methodology along with its validation reported by Tapase and Ranadive (2016, 2017), here at selected trail temperatures as per IRC:37-2012 and Indian conditions, the critical performance parameters i.e. the horizontal tensile strain at bottom of bituminous layer and vertical compressive strain at top of subgrade is examined with a program coded in FORTRAN and checked against the results obtained through the general purpose software ANSYS with plane 42 element for the axi-symmetric two-dimensional analysis.

4 Results and Conclusions

The critical performance parameters like horizontal tensile strain at the bottom of bituminous layer and vertical compressive strain at top of subgrade are examined for variation in the stiffness of materials in a bituminous layer of flexible pavement for selected trail temperatures. The bituminous mixes, as reported in Table 1 are considered for analysis and its effect on horizontal tensile strain at the bottom of the bituminous layer for various hypothetical temperatures are also considered for analysis. From the obtained results it is noticed that the value of horizontal tensile strain increases at each trial increase in temperature, showing the adverse effect of a rise in temperature on the life of the pavement. When the stiffer binder is used considering the decrement in temperature of the underneath layer, a drastic increase in the value of tensile strain is noticed which is at much higher side than the allowable limits (Fig. 3). The observed value indicates that the selected combination will give rise to fatigue cracking. On the other hand negligible variation in vertical compressive strain (Fig. 2) is observed. When the increment in temperature is considered then the reduction in the value of tensile strain even up to 35% is noticed. From the analysis, it is concluded that the use of too soft bituminous mix results in lowering the structural ability of the pavement at high temperatures and too hard bituminous mixes would become brittle at low temperatures resulting in cracking under loading.

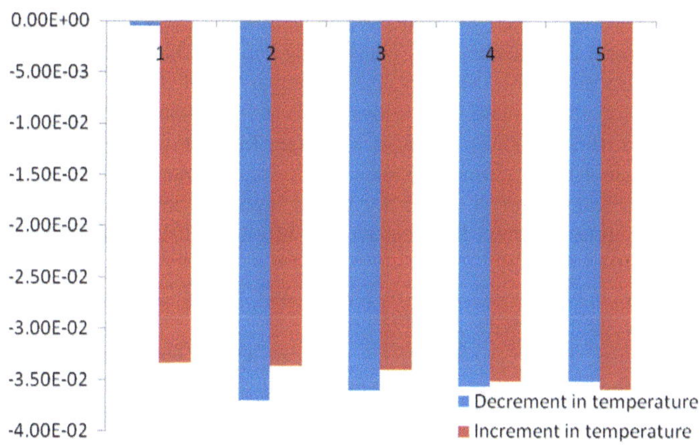

Fig. 2. Variation in vertical compressive strain at increment and decrement in temperature profile

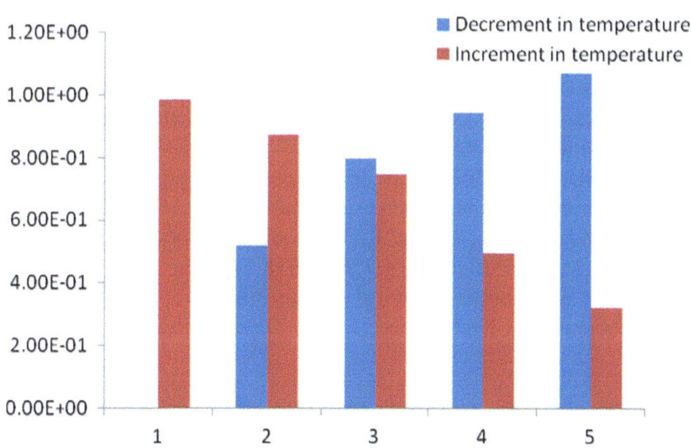

Fig. 3. Variation in Horizontal tensile strain at increment and decrement in temperature profile

References

American Association of State Highway Officials AASHTO: AASHTO Guide for the Design of Pavement Structures. AASHTO, Washington, D.C. (1993)

Animesh, D., Pandey, B.B.: The m-e design of bituminous road: an Indian perspective. J. Transp. Eng ASCE **125**(5), 463–471 (1999). https://doi.org/10.1061/(ASCE)0733-947X(1999)125:5(463)

Chakraborty, P., Das, A.: Principles of Transportation Engineering. Prentice Hall of India private limited (2003)

Chandak, P.G., Tapase, A.B., Sayyed, S.S., Attar, A.C.: A state-of-the-art review of different conditions influencing the behavioral aspects of flexible pavement. Advancement in the Design and Performance of Sustainable Asphalt Pavements, pp. 300–312. Springer International Publishing, Berlin (2018). https://doi.org/10.1007/978-3-319-61908-8_22

Chiasson, A., Yavuzturk, C., Ksaibati, K.: Linearized approach for predicting thermal stresses in asphalt pavements due to environmental conditions. J. Mater. Civ. Eng. ASCE **20**(2), 118–127 (2008). https://doi.org/10.1061/(asce)0899-1561

Diefenderfer, B.K., Al-Qadi, I.L., Diefenderfer, S.D.: Model to predict pavement temperature profile: development and validation. J. Transp. Eng. ASCE **132**(2), 162–167 (2006)

El-shaib, M.A., El-Badawy, S.M., Shawaly, E.A.: Comparison of AASHTO 1993 and MEPDG considering the Egyptian climatic conditions. Innovative Infrastruct. Solution, **2**(18) (2017). https://doi.org/10.1007/s41062-017-0067-6. (Springer International Publishing, Switzerland)

Ghosha, A., Padmarekhab, A., Murali Krishnan, J.: Implementation and proof-checking of mechanistic-empirical pavement design for indian highways using AASHTOWARE pavement me design software. Proc. Soc. Behav. Sci. 119–128 (2013). https://doi.org/10.1016/j.sbspro.2013.11.104

Gogoi, R., Das, A., Chakraborty, P.: Are fatigue and rutting distress modes related? Int. J. Pavement Res. Technol. **6**(4), 269–273 (2013)

Hasan, M.R.M., Hiller, J.E., You, Z.: Effects of mean annual temperature and mean annual precipitation on the performance of flexible pavement using ME design. Int. J. Pavement Eng. (2015). https://doi.org/10.1080/10298436.2015.1019504

Helwany, S., Dyer, J., Leidy, J.: Finite element analysis of flexible pavement. J. Transp. Eng. ASCE, **124**(5), 491–499 (1998). https://doi.org/10.1061/(asce)0733947X(1998)124:5(491)

Immanuel, S., Timm, D.H.: Measured and theoretical pressures in base and subgrade layers under dynamic truck loading. In: Proceeding, 2006 Airfield and Highway Pavement Specialty Conference, American Society of Civil Engineers, pp. 155–166 (2006)

IRC: 37–2012: Guidelines for the Design of Flexible Pavements. Indian Roads Congress, New Delhi (2012)

Maadani, O., Abd El Halim, A.O.: Environmental considerations in the AASHTO mechanistic-empirical pavement design guide: impacts on performance. J. Cold Reg. Eng. ASCE (2017). ISSN 0887-381X

Masad, S., Little, D., Masad, E.: Analysis of flexible pavement response and performance using isotropic and anisotropic material properties. J. Transp. Eng. ASCE **132**(4), 342–349 (2006)

Mulungye, R.M., Owende, P.M.O., Mellon, K.: Finite element modeling of flexible pavement on soft soil subgrades. Mater. Des. **28**, 739–756 (2007). Elsevier

NCHRP: Evaluation of Mechanistic-Empirical Design Procedure. National Cooperative Highway Research Program, NCHRP Project 1-37A, National Research Council, Washington (2007)

Orobioa, A., Zaniewskib, J.P.: Sensitivity of the mechanistic-empirical pavement design guide to traffic inputs: a space-filling approach. Road Mater. Pavement Des **14**(3), 735–746 (2013). https://doi.org/10.1080/14680629.2013.816248. Taylor & Francis

Rahman, M.T., Mahmud, K., Ahsan, S.: Stress-Strain characteristics of flexible pavement using finite element analysis. Int. J. Civ. Struct. Eng. **2**(1), 233–240 (2011)

Ranadive, M.S., Katkar, A.B.: Finite element analysis of flexible pavements. Indian High. **38**(6) (2010)

Ranadive, M.S., Tapase, A.B.: Parameter sensitive analysis of flexible pavement. Int. J. Pavement Res. Technol. (IJPRT). Elsevier, Special Issue on Sustainability on Pavement Engineering (2016a). https://doi.org/10.1016/j.ijprt.2016.12.001

Ranadive, M.S., Tapase, A.B.: Pavement performance evaluation for different combinations of temperature conditions and bituminous mixes. Innovative Infrastruct. Solutions **1**(40) (2016b). https://doi.org/10.1007/s41062-016-0040-9

Ranadive, M.S., Tapase, A.B.: Investigation of behavioral aspects of flexible pavement under various conditions by finite element method. Constitutive Model Geomater (2013). https://doi.org/10.1007/978-3-642-32814-5_100

Sahaa, J., Nassirib, S., Bayatc, A., Soleymanid, H.: Evaluation of the effects of Canadian climate conditions on the MEPDG predictions for flexible pavement performance. Int. J. Pavement Eng. **15**(3) (2014). https://doi.org/10.1080/10298436.2012.752488

Siddharthan, R.V., Sebaaly, P.E., El-Desouky, M., Strand, D., Huft, David: Heavy off-road vehicle tyre-pavement interaction and response. J. Transp. Eng. ASCE **131**(3), 239–247 (2005). https://doi.org/10.1061/(ASCE)0733-947X(2005)131:3(239)

Sinha, A.K., Chandra, S., Kumar, P.: Finite element analysis of flexible pavement with different subbase materials. Indian Highw. New Delhi **42**(2), 53–63 (2014)

Su, K., Sun, L., Hachiya, Y., Maekawa, R.: Analysis of shear stress in asphalt pavement contact pressure. In: 6th ICPT Sapporo, Japan (2008)

Tapase, A., Ranadive, M.: Performance evaluation of flexible pavement using finite element method. In: Material, Design, Construction, Maintenance and testing of pavement (2016). https://doi.org/10.1061/9780784480090.002

Tapase A.B., Ranadive, M.S.: Predicting performance of flexible pavement using finite element method. Advancement in the design and performance of sustainable asphalt pavements. Sustain. Civ. Infrastruct. (2017). https://doi.org/10.1007/978-3-319-61908-8_11

Wang, G., Roque, R.: Evaluation of truck tyre types on near-surface pavement response based on finite element analysis. Int. J. Pavement Res. Technol. **4**(4), 203–211 (2011)

Wang, Y., Chong, D.: Long-life flexible pavement: myth, reality, and the way forward. In: ASCE New Frontiers in Road and Airport Engineering, pp. 268–283 (2015)

Yang, H.H.: Pavement Analysis and Design, 2nd edn. Pearson Education Inc, and Dorling Kindersley Publishing, Inc (2008)

Yang, X., You, Z., Hiller, J., Watkins, D.: Sensitivity of flexible pavement design to Michigan's climatic inputs using pavement ME design. Int. J. Pavement Eng. (2015). https://doi.org/10.1080/10298436.2015.1105373. Taylor & Francis

Yoder, E.J., Witczak, M.W.: Principles of Pavement Design, 2nd edn. Wiley-Interscience Publication (1975)

Zeevaert, L.: Foundation Engineering for Difficult Subsoil Conditions, 2nd edn. Van Nostrand Reinhold company Inc (1982)

Zienkiewicz, O.C., Taylor, R.L.: The finite element method, vol. 2. McGraw Hill, New York (1991)

Zuo, G., Drumm, E.C., Meier, R.W.: Environmental effects on the predicted service life of flexible pavements. J. Transp. Eng. **133**(1), 47–56 (2007)

Automated Acoustic Scanning of Concrete Bridge Decks and Delamination Identification

Hongbin Sun and Jinying Zhu(✉)

Department of Civil Engineering, University of Nebraska-Lincoln, Omaha, USA
{hbsun, jyzhu}@unl.edu

Abstract. Chain drag testing is a common practice for bridge deck evaluation owing to its low cost and ease of use. Delaminations in bridge decks are identified by hollow sounds when steel chain is dragged on deck surface. However, this method is subjective, and result interpretation depends on the experience of the operators. Traffic noise also affects the test speed and accuracy of results. This paper presents an automated acoustic scanning system to detect delaminations in concrete structures, including bridge decks. The system consists of an array of ball-chain impactors, a noncontact MEMS microphone sensor array, multi-channel data acquisition device, RTK GPS positioning system, and signal processing schemes. A delamination identification algorithm was then developed to identify the positions and dimensions of delaminations. The acoustic scanning system and the delamination identification algorithm were validated in field tests of concrete bridge decks. The system can scan a concrete bridge deck at walking speed with half- or full-lane width coverage. Compared to the conventional manual chain drag test, the automated system provides improved efficiency, accuracy, repeatability, and practicality.

1 Introduction

Concrete bridge decks experience deterioration over time due to various types of aging effects. Delaminations caused by concrete deterioration and rebar corrosion are the most common defect in concrete bridge decks. Chain drag or hammer sounding is common practice for delamination evaluation due to its low cost and ease of use. A major drawback of chain drag is that it relies on subjective interpretation of the inspector. The Impact-echo (IE) is a proven nondestructive testing (NDT) method for concrete evaluation (Cheng and Sansalone 1993; Sansalone and Streett 1997). The IE test not only identifies delaminations but also gives the depth of delamination. However, the IE test requires contact between sensor/source and concrete surface, which makes it time-consuming and labor intensive. Zhu and Popovics (2007) proposed air-coupled IE test by using a microphone to replace the contact sensor, which significantly improves the test speed. Base on their pioneering study, researchers have attempted to combine automated impact source and air-coupled sensing to increase the test efficiency (Gucunski et al. 2012; Kee and Gucunski 2016; Mazzeo et al. 2016; Zhang et al. 2012). However, these systems need complex electrical and mechanical control for consistent impacts and the contact impact source remains a challenge for rapid scanning. On the other hand, the chain drag test has an advantage of using a

D. Chen et al. (eds.), *Advancements on Sustainable Civil Infrastructures*,
Sustainable Civil Infrastructures, https://doi.org/10.1007/978-3-319-96241-2_13

simple acoustic source for continuous excitation. Costley et al. (2003), Costley and Boudreaux (2003) developed an automated chain dragging system that includes dragging chains, a microphone, and signal acquisition and processing components. If working properly, the system will be more efficient and consistent than the manual chain drag test. However, we found that the chains generated broadband noises that may lead to misjudgment in some cases.

In this study, a multi-channel acoustic scanning system is presented which integrates continuous impact excitation with the newly developed ball chains, MEMs microphones, and a high accuracy RTK GPS positioning system. The system enables automated imaging of structures for rapid mapping delaminations in bridge decks with high spatial resolution. The ball-chain impactors allow impact-type acoustic excitations in a continual manner. A post processing algorithm is developed to identify and characterize delaminations. Field test on two different kinds of bridge decks were conducted to validate the repeatability and accuracy of the acoustic scanning system.

2 Acoustic Testing System

The acoustic scanning system includes excitation sources (chains), acoustic sensors (microphones), and positioning devices. Low cost MEMs (Micro-electro-mechanical systems) microphones (Adafruit SPW2430) with a frequency range of 100 Hz–10 kHz were used as acoustic sensors. Microphones and chains were installed on a scanning frame. The testing frame includes 12 channels, and each channel includes one MEMs microphone and one chain installed below each microphone. The channel spacing is about 0.15 m (6 in.) which determines the lateral spatial resolution of scanning. Acoustic signals received by the microphone array were acquired in continuous streaming mode by two oscilloscopes, PicoScope 4824 (8 channels) and PicoScope 5444a (4 channels), with a sampling rate of 100 kHz. The positioning data was also synchronized and recorded. A LabVIEW program with user interface was developed to control data acquisition and analysis. A schematic diagram of the acoustic testing system is shown in Fig. 1.

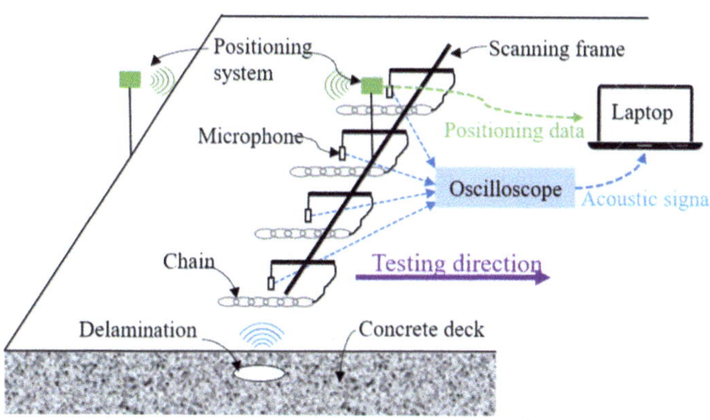

Fig. 1. Schematic diagram of the automated scanning system

A newly developed ball-chain is used in our system. To differentiate the conventional chains and the ball- chain impactors, we denote the conventional chain as "link-chain", and the new source as "ball-chain". Figure 2 shows two different types of steel link-chains (C1 and C2) and the newly developed ball-chain. Both steel link-chains C1 and C2 have a 6.35 mm (1/4 in.) diameter, while C2 has galvanized finish. The ball-chain consists of two 12.7 mm (1/2-in.) and two 15.9 mm (5/8-in.) diameter brass balls with a spacing of 2 cm. In our previous study (Sun et al. 2017), the ball-chain shows higher signal-to-noise ratio (S/N) and is more sensitive to delaminations than the traditional link-chain.

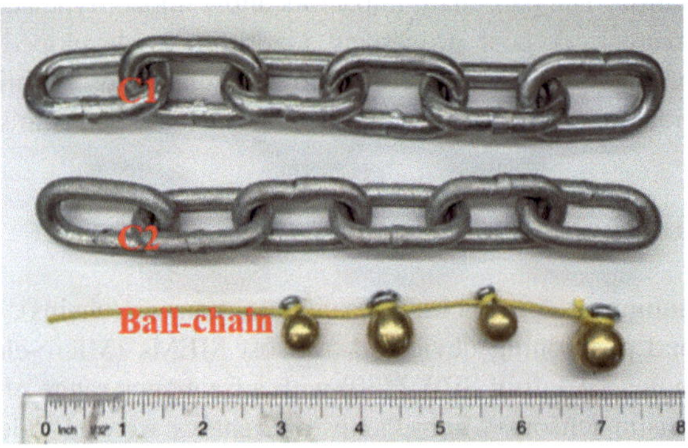

Fig. 2. Two steel link-chains and a ball-chain

A Real Time Kinematic (RTK) GPS system (Piksi GPS, Swift Navigation, Inc.) provided real-time positioning during the scanning. Before the field test, the RTK GPS accuracy was checked at a concrete parking lot with joints in south-north and west-east directions. These joints (grid lines in Fig. 3) were used as the reference to check the GPS rover moving paths. The base unit was placed at the origin in Fig. 3 and the rover unit moved from the start point along the joints. The moving paths matched well with the joints. The RTK GPS accuracy was also validated in many field tests.

3 Signal Processing and Delamination Identification

After scanning, the recorded acoustic signals from the 12 channel microphones were processed by short-time Fourier transform (STFT). A time-frequency spectrogram image is generated from each signal. Then the STFT amplitudes in the frequency range of 0.5–5 kHz were summed up to form a one dimensional data set versus scanning time for each channel (Sun et al. 2017). By synchronizing the RTK GPS position with the acoustic signals, the scanning time axis in the spectrogram is replaced by the position coordinates at that time. Then the data sets from all 12 channels were stacked together to generate a two-dimensional map image with two axes representing dimensions in the longitudinal (scanning) and the transverse directions. The value of each pixel in the map represents the summed STFT amplitude in the frequency range 0.5–5 kHz, and a

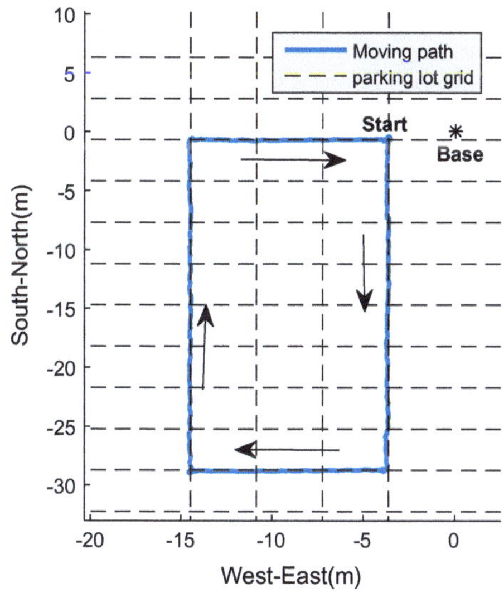

Fig. 3. Moving path of RTK GPS on the parking lot

large value represents a high delamination response. In this paper, red color pixels represent high signal amplitudes, while deep blue color represent low amplitudes. A schematic diagram is shown in Fig. 4 to demonstrate the signal processing.

A delamination identification algorithm was then developed to identify the locations and dimensions of the delaminations. A two-dimensional matrix P represents the pixel amplitudes of the final scanning image. According to the authors' previous study (Sun et al. 2017), acoustic signals in delamination areas have much higher amplitudes (summed STFT in 0.5–5 kHz) than in the sound areas. The algorithm can be described as the following steps:

(1) The matrix P is first normalized with respect to the maximum value in P, and then converted to a binary matrix using threshold P_t:

$$P(i,j) = \begin{cases} 0 \; if \; P(i,j) < P_t \\ 1 \; if \; P(i,j) \geq P_t \end{cases} \tag{1}$$

(2) Moore-Neighbor tracing algorithm modified by Jacob's stopping criteria is used to trace the boundary of each identified delamination (Gonzalez et al. 2004);
(3) An enclosing rectangular shape boundary is generated based on the maximum length and width of the traced boundary of each delamination.
(4) Location and dimensions of the delamination is determined from the rectangle.

A delamination detected in one field test is used to illustrate the boundary tracing process using the algorithm described above. In Fig. 5, the red area represents a detected delamination with high signal amplitude, and the blue color represents low

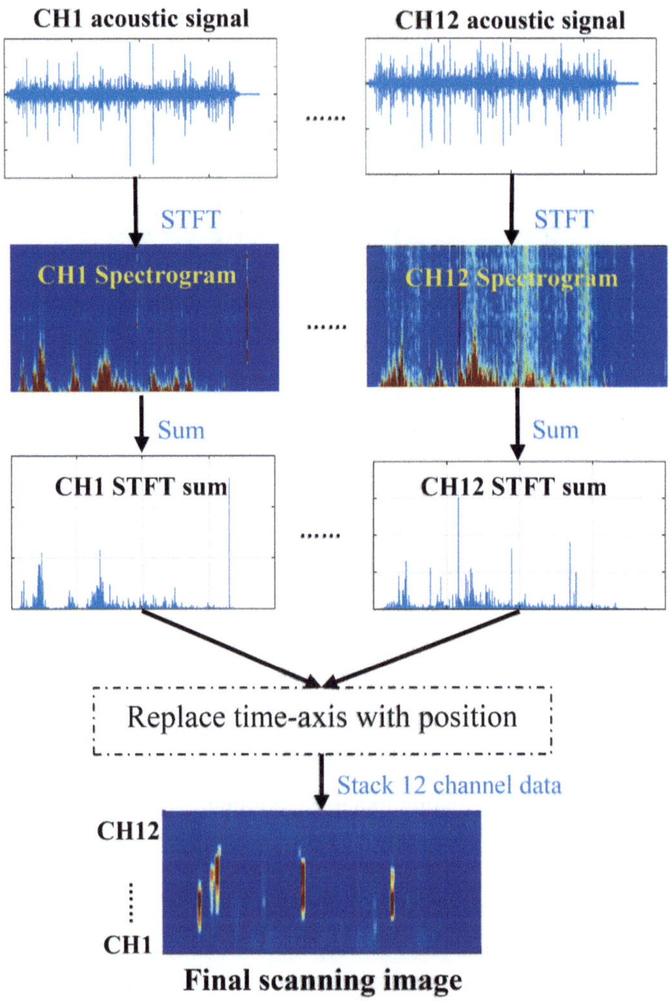

Fig. 4. Scheme diagram of the signal processing algorithm

amplitude of background signals in sound areas. Experimental study indicates that the normalized amplitudes of background signals in the blue areas are only about 1/4000. As a comparison, the amplitudes in the red area are near 1.0. It is found that using the threshold value $P_t = 0.1$ gives reasonable size of the delamination.

The boundary tracing results of this delamination is compared to the manual chain drag results. The traced boundary was acquired using the tracing algorithm and plotted in black dashed curve. The rectangular boundary (red rectangle) encloses the traced boundary with the maximum length (transverse direction) and width (longitudinal direction) of the traced boundary. The width of delamination identified by the automated algorithm tends to be smaller than the manual chain drag results, while the length is similar to the manual results.

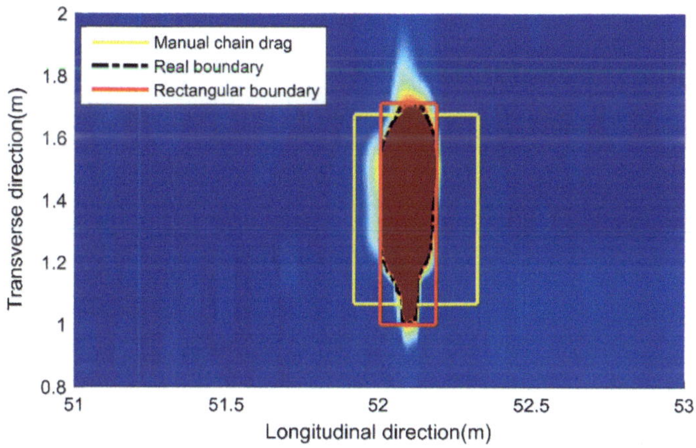

Fig. 5. Delamination boundary tracing of Delamination #1

4 Field Testing on Bridge Decks

4.1 Scanning Results on Two Bridge Decks

The scanning system was tested on two different types of bridge decks in Nebraska, USA. The first bridge had an asphalt overlaid concrete deck, and acoustic scanning was performed after removal of asphalt overlay. The second bridge had a bare concrete deck with transverse tining. The acoustic scanning results are presented below.

4.1.1 Asphalt Overlaid Concrete Deck

A concrete bridge deck with asphalt overlay in Nebraska was undergoing repair in July 2017. The asphalt overlay was first removed using a milling machine to expose the concrete deck surface. The exposed concrete deck has a very rough surface, which may present challenges to other NDT methods, such as the impact-echo test. In this study, the deck was scanned using the automated acoustic scanning system and manual chain drag. Only the westbound lane and the shoulder (yellow area in Fig. 6a) were tested. Figure 6b shows the deck surface conditions after asphalt removal.

Figure 7 shows the delamination map of the scanned concrete deck in Fig. 6a. The red areas represent delaminations and the white rectangles are traced delamination boundaries using the boundary tracing algorithm. The center positions and the sizes of the delaminations were determined from the rectangles. In Fig. 7, 33 delaminations were identified and the total delamination area was 17.2 m^2 (scanned area was 61.56 m^2). The delamination area percentage is the ratio between the delamination area and the total scanned area, which is an important factor for maintenance decision making. For this bridge deck, the delamination percentage was about 28%. It is also noticed that most delaminations are located in the traffic lane along two wheel paths, which experienced heavier traffic loads than the bridge shoulder and the area between wheel paths.

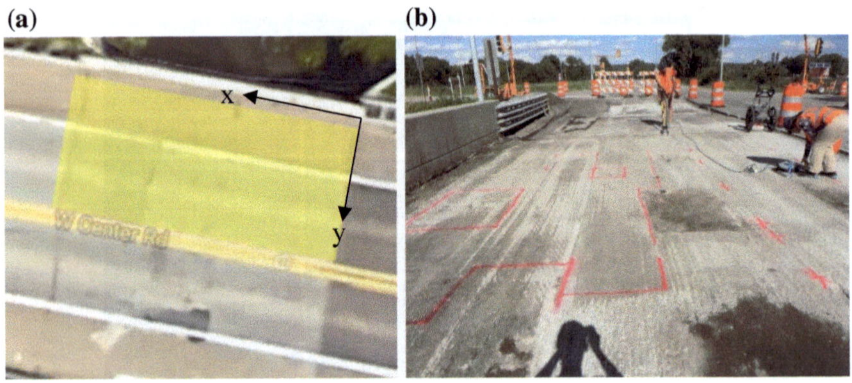

Fig. 6. **a** Google map of the bridge, **b** concrete deck after asphalt removal

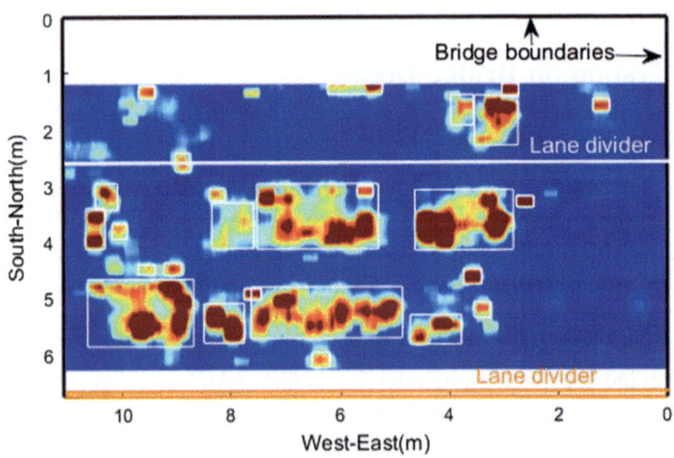

Fig. 7. Delamination map of the scanning results

Manual chain drag test was also used to directly mark delaminations on deck surface. Since this bridge deck had wide spread severe delamination, it was difficult for manual chain drag test to clearly identify delamination boundaries. Therefore, the delamination areas marked by manual chain drag is much larger than the areas identified by the automated scanning system.

4.1.2 A 95 m Long Bare Concrete Decks

A 95 m long concrete bridge deck in south Lincoln Nebraska was tested in 2017 using the automated acoustic scanning system. Manual chain drag test was also conducted, and the detected delaminations were marked on the deck. The manual chain drag results were used as the verification of the acoustic scanning results (Fig. 8).

Figure 9 shows the acoustic scanning images of the 95-m long bridge deck. For better resolution, the image is split into two parts (0–45 and 45–95 m). In the images,

Fig. 8. Google map of the 95-m long bridge deck (Lincoln, Nebraska) and the scanned area

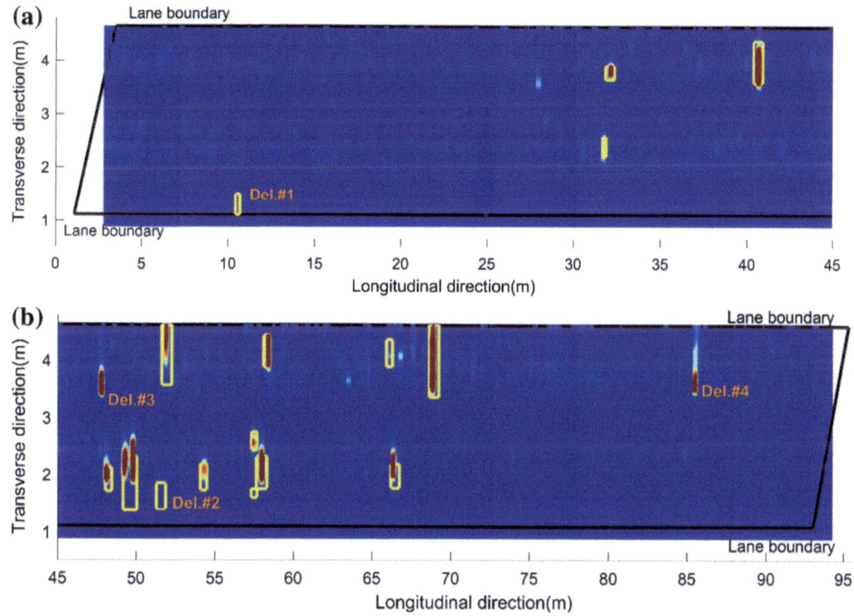

Fig. 9. Scanning images of the 95-m long bridge deck: **a** 0–45 m, **b** 45–95 m

black lines represent the bridge lane boundaries. Red areas represent the delaminations and the blue areas represent the sound deck areas. Delaminations identified by the manual chain drag are shown as yellow rectangles. It can be seen that the automated scanning results agree well with the manual chain drag results, except for a few small delaminations. For example, Del. #1 and Del. #2 were identified by the manual chain drag, but they were missed by the automated scanning. It could be because no ball chain impacted at those locations during the scan. On the other hand, the automated scanning found two additional delaminations (Del. #3 and Del. #4) that were missed by the manual chain drag. These delaminations were confirmed by hammer sounding and multiple scans.

4.2 Testing Repeatability

To investigate the repeatability of the acoustic scanning system, four scans were collected on the bare concrete bridge deck shown in Fig. 8. scan #1 and scan #3 were tested along the same direction and scan #2 and scan #4 were from the opposite direction. In Fig. 10, a scanned area of 30 m (100 ft) long by 1.75 m wide with several delaminations was chosen to compare the four scanning results.

By comparing the results of four scans, we find all four images agree well with each other in terms of the locations, shapes and dimensions of all major delaminations. There are some minor differences for small size delaminations, such as the delamination near 55 m position, among the four results. scan #3 and scan #4 clearly show this delamination with high amplitudes, while scans #1 and #2 show it with lower amplitude. In scan #2, there is a major delamination at 58 m and two small delaminations at both ends of the major delamination. These two small delaminations are also presented in scan #4 with lower amplitude. However, they are not shown in scans #1 and #3 images. The manual chain drag confirmed these two small delaminations. The results suggest scanning direction may have some effects on small delamination identification. Therefore, multiple scans from opposite directions are suggested to provide more accurate evaluation of bridge deck conditions.

5 Conclusions

In this study, an automated acoustic system was developed by using a ball-chain source for rapid bridge deck evaluation. A signal processing algorithm and an image processing algorithm were proposed for delamination identification and characterization (position and size). The system was validated in multiple field tests on concrete bridge decks. Based on the results in this paper, the following conclusions are drawn:

1. The acoustic scanning system provides rapid and reliable evaluation of delaminations in concrete bridge decks.
2. The system is robust even on rough deck surfaces.
3. The automated system shows good repeatability. Multiple scans along different scanning directions will further improve the test reliability and reduce the chance of missed defects.
4. The RTK GPS provides high precision positioning data (1 cm accuracy), which can be synchronized with acoustic signals to provide real time test positions.
5. The delamination identification algorithm is effective to identify and characterize delaminations automatically.

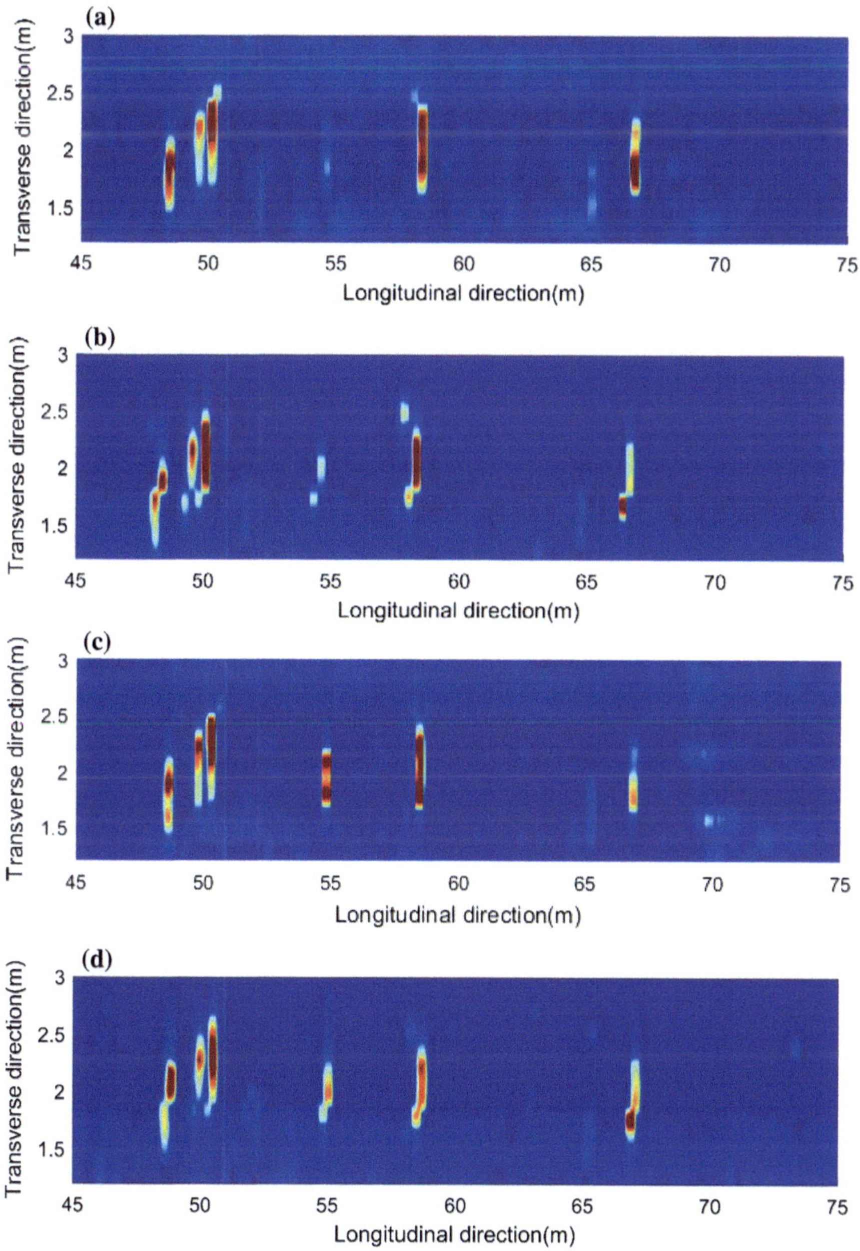

Fig. 10. Scanning results of a 30 m long bridge deck section: **a** scan #1, **b** scan #2, **c** scan #3, **d** scan #4

Acknowledgements. This study was sponsored by the Nebraska Department of Transportation (NDOT). The Bridge Division of NDOT is gratefully acknowledged for their support on traffic control during the field test. The authors would also like to thank Sepehr Pashoutani and Hossein Arian for their help on the field test.

References

Cheng, C., Sansalone, M.: The impact-echo response of concrete plates containing delaminations: numerical, experimental and field studies. Mater. Struct. **26**(5), 274–285 (1993). Springer

Costley, R.D., Boudreaux, G.M.: Finding delaminations in concrete bridge decks. 146th ASA Meeting, Austin, TX, Austin, TX (2003)

Costley, R.D., Henderson, M.E., Dion, G.N.: Acoustic Inspection of Structures. United States Patent and Trademark Office, US (2003)

Gonzalez, R.C., Woods, R.E., Eddins, S.L.: Digital Image Using Matlab Processing. Lexington, Person Prentice Hall (2004)

Gucunski, N., Yan, M., Wang, Z., Fang, T., Maher, A.: Rapid bridge deck condition assessment using three-dimensional visualization of impact echo data. J. Infrastruct. Syst. **18**(1), 12–24 (2012)

Kee, S., Gucunski, N.: Interpretation of flexural vibration modes from impact-echo testing. J. Infrastruct. Syst. **22**(3), 1–10 (2016)

Mazzeo, B.A., Larsen, J., McElderry, J., Guthrie, W.S.: Rapid multichannel impact-echo scanning of concrete bridge decks from a continuously moving platform. Review of Progress in Quantitative Nondestructive Evaluation 2016 (QNDE 2016), Atlanta, USA (2016)

Sansalone, M.J., Streett, W.B.: Impact-Echo. Nondestructive Evaluation of Concrete and Masonry. Bullbrier Press, Jersey Shore, PA (1997)

Sun, H., Zhu, J., Ham, S.H.: Acoustic evaluation of concrete delaminations using ball-chain impact excitation. J. Acoust. Soc. Am. Express Lett. **141**(5), 1–11 (2017)

Zhang, G., Harichandran, R.S., Ramuhalli, P.: An automatic impact-based delamination detection system for concrete bridge decks. NDT & E Int. **45**(1), 120–127 (2012). Elsevier

Zhu, J., Popovics, J.S.: Imaging concrete structures using air-coupled impact-echo. J. Engi. Mech. Am. Soc. Civ. Eng. **133**(6), 628–640 (2007)

Author Index

D. Chen et al. (eds.), *Advancements on Sustainable Civil Infrastructures*,
Sustainable Civil Infrastructures, https://doi.org/10.1007/978-3-319-96241-2

MIX
Papier aus verantwortungsvollen Quellen
Paper from responsible sources
FSC® C105338

If you have any concerns about our products,
you can contact us on
ProductSafety@springernature.com

In case Publisher is established outside the EU,
the EU authorized representative is:
Springer Nature Customer Service Center GmbH
Europaplatz 3, 69115 Heidelberg, Germany

Printed by Libri Plureos GmbH
in Hamburg, Germany